D1203747

A Collection of Matrices
for Testing Computational Algorithms

A Collection of Matrices for Testing Computational Algorithms

ROBERT T. GREGORY, Ph.D.
Professor of Mathematics and of Computer Sciences
Senior Research Mathematician, Computation Center
The University of Texas at Austin

DAVID L. KARNEY, M.A.
Department of Computer Sciences
The University of Texas at Austin

WILEY-INTERSCIENCE, a division of John Wiley & Sons
New York • London • Sydney • Toronto

Library of Congress Catalogue Card Number: 70-95578

SBN 471 32669 0

Printed in the United States of America

10 9 8 7 6 5 4 3 2 1

To Alston S. Householder

PREFACE

This monograph is intended primarily as a reference book for numerical analysts and others who are interested in computational methods for solving problems in matrix algebra. It is well known that a good mathematical algorithm may or may not be a good computational algorithm. Consequently, what is needed is a collection of numerical examples with which to test each algorithm as soon as it is proposed. It is our hope that the matrices we have collected will help fulfill this need.

The test matrices in this collection were obtained for the most part by searching the current literature. However, four individuals who had begun collections of their own contributed greatly to this effort by providing a large number of test matrices at one time.

First, Joseph Elliott's Master's thesis [18] provided a large collection of tridiagonal matrices. Second, Mrs. Susan Voigt, of the Naval Ship Research and Development Center, contributed a varied collection of matrices. Third, Professor Robert E. Greenwood, of The University of Texas at Austin, provided a valuable list of references along with his collection of matrices and determinants. Finally, just as this work was nearing completion, the collection of Dr. Joan Westlake [60] was discovered. Her collection of 41 test matrices contained seven which we had overlooked; therefore, they were added.

Matrix 6.11 is a non-Hermitian matrix of order 20. It is a specific example of a class of matrices known as Dolph-Lewis matrices [14] which arose around 1957 in an investigation of perturbations of plane Poiseuille flow. Accurate eigenvalues, along with left and right eigenvectors and condition numbers, were provided by Dr. J. H. Wilkinson of the National Physical Laboratory.

Matrix 3.8 is the finite segment (of order n) of the (infinite) Hilbert Matrix. Matrix 3.26 is a generalization. The exact inverses of the finite Hilbert segments exhibited were provided by Dr. Max Engeli of FIDES Treuhand-Vereinigung, Zürich. Dr. Engeli's program for computing these inverses was written in SYMBAL, a language of his own creation.

The first author is grateful to Dr. Engeli and to Dr. Erwin Nievergelt for making the facilities of FIDES, including the CDC 6500 computer, available to him during his stay in Zürich.

Partial support for this work was provided by the National Science Foundation under Grant GP 8442 and by the Army Research Office (Durham) under Grant DA-ARO(D)-31-124-G721, at The University of Texas at Austin. This support is gratefully acknowledged. We are also grateful to Professor George E. Forsythe, who read the manuscript and offered many helpful suggestions.

The book is dedicated to Dr. Alston S. Householder, who has inspired numerical analysts for the past two decades.

We are indebted to Mrs. Dorothy Baker for preparing the manuscript. Her superb job of typing this difficult material enabled the publishers to use photographic reproduction. This saved the authors an enormous amount of additional proofreading and avoided the introduction of countless additional errors.

ROBERT T. GREGORY
DAVID L. KARNEY

AUSTIN, TEXAS
April 1969

TABLE OF CONTENTS

CHAPTER I

INTRODUCTION

In order to test the accuracy of computer programs for solving numerical problems, one needs numerical examples with known solutions. The aim of this monograph is to provide the reader with suitable examples for testing algorithms for finding the inverses, eigenvalues, and eigenvectors of matrices. A collection of methods for constructing test matrices and a large collection of numerical examples have been included. We have endeavored to allow the reader much freedom in his choice of a test matrix.

Chapter II of this monograph describes methods for generating matrices with known inverses and eigensystems whereas Chapter III contains test matrices with known inverses and solutions of systems of linear equations.

In the later chapters test matrices with known eigenvalues and eigenvectors are given. We have included, when possible, both right and left eigenvectors. The reader is reminded that if A is an Hermitian matrix, the left eigenvectors of A are the conjugate transpose of the right eigenvectors. For some of the examples, the tridiagonal forms are given which arise in the use of certain well-known algorithms for computing eigenvalues. The methods of Givens and Householder, for example, transform real symmetric matrices into the tridiagonal form

$$\begin{bmatrix} \alpha_1 & \beta_2 & & & \\ \beta_2 & \alpha_2 & \beta_3 & & \\ & \cdots & \cdots & \cdots & \\ & & \beta_{n-1} & \alpha_{n-1} & \beta_n \\ & & & \beta_n & \alpha_n \end{bmatrix} .$$

The method of Lanczos transforms nonsymmetric matrices into the tridiagonal form

$$\begin{bmatrix} \alpha_1 & \beta_2 & & & \\ 1 & \alpha_2 & \beta_3 & & \\ & \cdots & \cdots & \cdots & \\ & & 1 & \alpha_{n-1} & \beta_n \\ & & & 1 & \alpha_n \end{bmatrix} .$$

The examples exhibited in this monograph include both well-conditioned and ill-conditioned matrices. For each example we have computed several condition numbers, and for the ill-conditioned matrices the condition numbers are included.

Let $A = [a_{ij}]$ be an n x n nonsingular matrix with eigenvalues $\lambda_1, \lambda_2, \ldots, \lambda_n$. For the problem of matrix inversion, at least three condition numbers are used. Von Neumann and Goldstine [59] suggest the condition number

$$P(A) = \frac{\max_i |\lambda_i|}{\min_i |\lambda_i|} ,$$

Turing [57] proposes the two condition numbers

$$M(A) = n \max_{i,j} |a_{ij}| \max_{i,j} |\alpha_{ij}|$$

and

$$N(A) = \frac{1}{n} \|A\|_E \cdot \|A^{-1}\|_E ,$$

where

$$\|A\|_E = \left[\sum_{i=1}^{n} \sum_{j=1}^{n} a_{ij}^2 \right]^{\frac{1}{2}} ,$$

and where

$$A^{-1} = [\alpha_{ij}].$$

It can be shown that P(A) and N(A) do not differ very much from M(A). In particular, we have [60, p. 90], [53]

$$\frac{1}{n^2} M(A) \leqq N(A) \leqq M(A)$$

and

$$P(A) \leqq nM(A).$$

If A is symmetric, we also have

$$\frac{1}{n} M(A) \leqq P(A).$$

If the matrix elements are chosen at random from a normal population, then an N-condition number of order $\sqrt{n}$ and an M-condition number of order $\sqrt{n} \log n$ can be expected.

Actually, M(A) and N(A) are not used as much as the more general condition number

$$K(A) = \|A\| \cdot \|A^{-1}\|,$$

for various norms, not necessarily the same.

Now let $x^{(i)}$ and $y^{(i)}$ be, respectively, right and left eigenvectors of A corresponding to the eigenvalue λ_i, and suppose $x^{(i)}$ and $y^{(i)}$ are normalized so that

$$\sum_{j=1}^{n} |x_j^{(i)}|^2 = \sum_{j=1}^{n} |y_j^{(i)}|^2 = 1.$$

The condition of A with respect to the eigenvalue problem can be measured by the n condition numbers of A [62, pp. 88-89],

$$s_i = \left(y^{(i)}, x^{(i)}\right), \quad i = 1, 2, \ldots, n.$$

Here, s_i is the condition number for λ_i. Thus, some eigenvalues may be more ill-conditioned than others. Observe that if A is Hermitian, we have, for all i,

$$s_i = 1.$$

CHAPTER II

CONSTRUCTION OF TEST MATRICES

1. In this chapter we present a variety of methods by which the reader can construct matrices with known inverses, eigenvalues, and eigenvectors. We begin with the following well-known results which can be found in elementary texts on matrix algebra such as Hohn [28].

Theorem 1. The eigenvalues of A and A^T are the same.

Theorem 2. The eigenvalues of $\bar{A}$ and A^H are the conjugates of the eigenvalues of A.

Theorem 3. The eigenvalues of A^{-1} are the inverses of the eigenvalues of A.

Theorem 4. If $\lambda_1, \lambda_2, \ldots, \lambda_n$ are the eigenvalues of an n x n matrix A and if $P(\alpha)$ is a polynomial, then the eigenvalues of P(A) are $P(\lambda_1), P(\lambda_2), \ldots, P(\lambda_n)$. Further, if x is an eigenvector of A corresponding to the eigenvalue λ, then x is an eigenvector of P(A) corresponding to $P(\lambda)$.

Theorem 5. The matrix

$$A = \begin{bmatrix} -a_1 & -a_2 & \cdots & -a_{n-1} & -a_n \\ 1 & 0 & \cdots & 0 & 0 \\ 0 & 1 & \cdots & 0 & 0 \\ \cdots & \cdots & \cdots & \cdots & \cdots \\ 0 & 0 & \cdots & 1 & 0 \end{bmatrix}$$

has the characteristic equation

$$|A-\lambda I| = \lambda^n + a_1\lambda^{n-1} + \dots + a_n = 0.$$

Theorem 6. If B is a non-singular matrix, then the eigenvalues of BAB^{-1} are the same as those of A. If x and y are, respectively, right and left eigenvectors of A corresponding to the eigenvalue λ, then Bx and yB^{-1} are respectively right and left eigenvectors of BAB^{-1} corresponding to the eigenvalue λ. If A is also non-singular, then $(BAB^{-1})^{-1} = BA^{-1}B^{-1}$.

2. One of the simplest methods of constructing test matrices is by forming composite matrices (some authors use compound matrices). In this regard we have the following.

Theorem 7. [2]. The eigenvalues of a block-diagonal matrix, diag $[A_1, A_2, \dots, A_k]$, are the eigenvalues of $A_1, A_2, \dots, A_k$.

Theorem 8 [28, pp. 81-82]. Suppose B is composed of submatrices of indicated orders,

$$B = \left[\begin{array}{c|c} A_{11} & A_{12} \\ (n \times n) & (n \times m) \\ \hline A_{21} & A_{22} \\ (m \times n) & (m \times m) \end{array}\right]$$

and suppose A_{11} and $P = A_{22} - A_{21}(A_{11}^{-1}A_{12})$ are non-singular. Then B is non-singular, and if we partition B^{-1} into submatrices

$$B^{-1} = \left[\begin{array}{c|c} B_{11} & B_{12} \\ (n \times n) & (n \times m) \\ \hline B_{21} & B_{22} \\ (m \times n) & (m \times m) \end{array}\right]$$

we have

$$B_{11} = A_{11}^{-1} + (A_{11}^{-1}A_{12})P^{-1}(A_{21}A_{11}^{-1})$$

$$B_{12} = -(A_{11}^{-1}A_{12})P^{-1}$$

$$B_{21} = -P^{-1}(A_{21}A_{11}^{-1})$$

$$B_{22} = P^{-1}.$$

Theorem 9 [36, p. 12]. If A and B are real n x n matrices and

$$S = \begin{bmatrix} A & B \\ B & A \end{bmatrix},$$

then the eigenvalues of S are the eigenvalues of A + B together with the eigenvalues of A - B.

Another class of composite matrices suitable for test purposes can be obtained by the use of Kronecker products. Most of the following material comes from Bellman [2, Chapter 12] and Marcus [36]; the reader is referred to Friedman [25] for additional information.

Definition [2]. Let $A = [a_{ij}]$ be an m x m matrix and B an n x n matrix. The mn x mn matrix defined by

$$\begin{bmatrix} a_{11}B & a_{12}B & \cdots & a_{1m}B \\ a_{21}B & a_{22}B & \cdots & a_{2m}B \\ \cdots & \cdots & \cdots & \cdots \\ a_{m1}B & a_{m2}B & \cdots & a_{mm}B \end{bmatrix}$$

is called the Kronecker product of A and B and is denoted by $A \otimes B$.

For matrices of this form we have the following very important results.

Theorem 10 [2]. If A is an m x m matrix with eigenvalues λ_i, i = 1, 2, ..., m, and B is an n x n matrix with eigenvalues μ_j, j = 1, 2, ..., n, then the eigenvalues of $A \otimes B$ are $\lambda_i \mu_j$, i = 1, 2, ..., m and j = 1, 2, ..., n. The eigenvectors are mn x 1 column-vectors of the form

$$z_{ij} = \begin{bmatrix} x_1^{(i)} y^{(j)} \\ x_2^{(i)} y^{(j)} \\ \vdots \\ x_m^{(i)} y^{(j)} \end{bmatrix}$$

where $y^{(j)}$ is an eigenvector of B corresponding to the eigenvalue μ_j and $x_k^{(i)}$, k = 1, 2, ..., m, denote the components of the eigenvector $x^{(i)}$ of A corresponding to λ_i.

Theorem 11 [36, p. 5]. If A and B are non-singular, then $A \otimes B$ is non-singular and $(A \otimes B)^{-1} = A^{-1} \otimes B^{-1}$.

We can, of course, consider Kronecker powers of a particular matrix, i.e.,

$$A^{(2)} = A \otimes A$$
$$A^{(k+1)} = A \otimes A^{(k)}.$$

The eigenvalues of $A^{(k)}$ are all possible products consisting of k factors, each of which is an eigenvalue of A. We can also define matrices having eigenvalues of this form which are of much smaller dimension than the general Kronecker product. For example [2], suppose

$$A = \begin{bmatrix} a_{11} & a_{12} \\ a_{21} & a_{22} \end{bmatrix}$$

and suppose A has eigenvalues λ_1, λ_2. Starting with the equations

$$\lambda_1 x_1 = a_{11}x_1 + a_{12}x_2$$
$$\lambda_1 x_2 = a_{21}x_1 + a_{22}x_2 ,$$

we form the products, for a fixed integer k,

$$(a_{11}x_1 + a_{12}x_2)^{k-i}(a_{21}x_1 + a_{22}x_2)^i, \quad i = 0, 1, \ldots, k.$$

Then, if we let $A_{(k)} = [b_{ij}]$, $i,j = 0, 1, \ldots, k$, denote the $k+1 \times k+1$ matrix such that b_{ij} is the coefficient of the $x_1^{k-j}x_2^j$ term in the product $(a_{11}x_1 + a_{12}x_2)^{k-i}(a_{21}x_1 + a_{22}x_2)^i$, the eigenvalues of $A_{(k)}$ are $\lambda_1^{k-i}\lambda_2^i$, $i = 0, 1, \ldots, k$.

For example, if k = 2, we have the products

$$(a_{11}x_1 + a_{12}x_2)^2 = a_{11}^2x_1^2 + 2a_{11}a_{12}x_1x_2 + a_{12}^2x_2^2$$

$$(a_{11}x_1 + a_{12}x_2)(a_{21}x_1 + a_{22}x_2) = a_{11}a_{21}x_1^2 + (a_{11}a_{22} + a_{12}a_{21})x_1x_2 + a_{12}a_{22}x_2^2$$

$$(a_{21}x_1 + a_{22}x_2)^2 = a_{21}^2x_1^2 + 2a_{21}a_{22}x_1x_2 + a_{22}^2x_2^2 .$$

Thus the matrix

$$A_{(2)} = \begin{bmatrix} a_{11}^2 & 2a_{11}a_{12} & a_{12}^2 \\ a_{11}a_{21} & (a_{11}a_{22}+a_{12}a_{21}) & a_{12}a_{22} \\ a_{21}^2 & 2a_{21}a_{22} & a_{22}^2 \end{bmatrix}$$

has eigenvalues $\lambda_1^2, \lambda_1\lambda_2, \lambda_2^2$.

Next let us suppose that A is an m x m matrix with eigenvalues λ_i, $i = 1, 2, \ldots, m$, and B is an n x n matrix with eigenvalues μ_j, $j = 1, 2, \ldots, n$. It can be shown that the mn eigenvalues of $(I_m \otimes B) + (A \otimes I_n)$ are $\lambda_i + \mu_j$, for all i and j. For example, let A be the m x m matrix

$$A = \begin{bmatrix} a & b & & & & \\ b & a & b & & & \\ & b & a & b & & \\ & \cdots & \cdots & \cdots & \cdots & \\ & & & b & a & b \\ & & & & b & a \end{bmatrix}$$

and let B be the n x n matrix

$$B = \begin{bmatrix} d & c & & & & \\ c & d & c & & & \\ & c & d & c & & \\ & \cdots & \cdots & \cdots & \cdots & \\ & & & c & d & c \\ & & & & c & d \end{bmatrix}$$

where the eigenvalues of A are $a + 2b \cos \frac{k\pi}{m+1}$, $k = 1, 2, \ldots, m$, and those of B are $d + 2c \cos \frac{k\pi}{n+1}$, $k = 1, 2, \ldots, n$. Then the eigenvalues of

$$(I_m \otimes B) + (A \otimes I_n) = \begin{bmatrix} (aI_n + B) & bI_n & & & \\ bI_n & (aI_n + B) & bI_n & & \\ & bI_n & (aI_n + B) & bI_n & \\ \cdots & \cdots & \cdots & \cdots & \cdots \\ & & & bI_n & (aI_n + B) \end{bmatrix}$$

are given by

$$\lambda_{ij} = a + d + 2b \cos \frac{i\pi}{m+1} + 2c \cos \frac{j\pi}{n+1},$$

$$i = 1, 2, \ldots, m; \; j = 1, 2, \ldots, n.$$

3. Determinants can also be used to define another type of matrix "power." For simplicity we shall consider a 3 x 3 matrix A and a set of 2 x 2 determinants formed from the eigenvectors of A. The procedure can be generalized to treat m x m determinants associated with the eigenvectors of n x n matrices [2].

Let

$$A = \begin{bmatrix} a \\ b \\ c \end{bmatrix} = \begin{bmatrix} a_1 & a_2 & a_3 \\ b_1 & b_2 & b_3 \\ c_1 & c_2 & c_3 \end{bmatrix}$$

and let the eigenvalues of A be λ_1, λ_2, λ_3 with associated eigenvectors $x^{(1)}$, $x^{(2)}$, $x^{(3)}$. Now define, for arbitrary n-dimensional vectors r and s,

$$g_{ij}(r,s) = \det \begin{bmatrix} r_i & s_i \\ r_j & s_j \end{bmatrix}, \quad i,j = 1, 2, \ldots, n.$$

Then, for our example, it can be shown that the matrix

$$G = \begin{bmatrix} g_{12}(a,b) & g_{23}(a,b) & g_{31}(a,b) \\ g_{12}(a,c) & g_{23}(a,c) & g_{31}(a,c) \\ g_{12}(b,c) & g_{23}(b,c) & g_{31}(b,c) \end{bmatrix}$$

has eigenvalues $\lambda_1\lambda_2$, $\lambda_1\lambda_3$, $\lambda_2\lambda_3$. Also, corresponding to the eigenvalue $\lambda_i\lambda_j$, $i \neq j$, there is an eigenvector

$$y = \begin{bmatrix} y_1 \\ y_2 \\ y_3 \end{bmatrix}$$

where

$$y_1 = g_{12}(x^{(i)},x^{(j)}), \; y_2 = g_{23}(x^{(i)},x^{(j)}), \text{ and } y_3 = g_{31}(x^{(i)},x^{(j)}).$$

4. Brenner [7] has described another set of composite matrices which can be used to test inversion and eigenvalue routines. Let f_n denote the $n \times 1$ column-vector whose components are all 1's. For arbitrary integers n and k, let $J_{nk} = f_n f_k^T$, i.e., J_{nk} is the $n \times k$ matrix whose elements are all 1's. The matrix J_{nn} has the following properties: f_n is an eigenvector of J_{nn} corresponding to the eigenvalue $\lambda = n$; every vector orthogonal to f_n is an eigenvector of J_{nn} corresponding to the eigenvalue $\lambda = 0$. The eigenvalue $\lambda = 0$ has multiplicity $n - 1$, and its associated invariant space is spanned by the vectors $g_i = f_n - n\, e_n^i$, $i = 1, 2, \ldots, n-1$, where e_n^i is the $n \times 1$ column vector which has components δ_{ij}, $j = 1, 2, \ldots, n$. This leads us to the following result.

Theorem 13. The matrix

$$A = \begin{bmatrix} (a_1 I_{n_1} + b_{11} J_{n_1 n_1}) & b_{12} J_{n_1 n_2} & \cdots & b_{1t} J_{n_1 n_t} \\ b_{21} J_{n_2 n_1} & (a_2 I_{n_2} + b_{22} J_{n_2 n_2}) & \cdots & b_{2t} J_{n_2 n_t} \\ \cdots & \cdots & \cdots & \cdots \\ b_{t1} J_{n_t n_1} & b_{t2} J_{n_t n_2} & \cdots & (a_t I_{n_t} + b_{tt} J_{n_t n_t}) \end{bmatrix}$$

is similar to the block-diagonal matrix

$$\text{diag}[A_1, A_2, \ldots, A_t, A_{t+1}]$$

where, for $i = 1, 2, \ldots, t$, $A_i = a_i I_{n_i - 1}$, and A_{t+1} is the $t \times t$ matrix defined by

$$A_{t+1} = \begin{bmatrix} (a_1 + b_{11} n_1) & b_{12} n_2 & \cdots & b_{1t} n_t \\ b_{21} n_1 & (a_2 + b_{22} n_2) & \cdots & b_{2t} n_t \\ \cdots & \cdots & \cdots & \cdots \\ b_{t1} n_1 & b_{t2} n_2 & \cdots & (a_t + b_{tt} n_t) \end{bmatrix}.$$

For $r = 1, 2, \ldots, t$, the vectors

$$x_{ir} = \begin{bmatrix} 0 \\ \vdots \\ 0 \\ f_{n_r} - n_r e^i_{n_r} \\ 0 \\ \vdots \\ 0 \end{bmatrix}, \quad i = 1, 2, \ldots, n_r - 1,$$

are the eigenvectors of A corresponding to the eigenvalue $\lambda = a_r$, which has multiplicity $n_r - 1$.

The determinants and inverses can also be obtained for matrices of this form. We illustrate with the example

$$B = \begin{bmatrix} (aI_n + bJ_{nn}) & cJ_{nm} \\ dJ_{mn} & (hI_m + kJ_{mm}) \end{bmatrix}.$$

The matrix B is similar to the matrix $\text{diag}[A_1, A_2, A_3]$ where

$$A_1 = aI_{n-1}$$

$$A_2 = hI_{m-1}$$

$$A_3 = \begin{bmatrix} (a+bn) & cm \\ dn & (h+km) \end{bmatrix}.$$

From this it follows that the eigenvalues of B are a, h, and the eigenvalues of A_3. Also, the determinant of B is seen to be $a^{n-1}h^{m-1}[(a+bn)(h+km)-cdnm]$. Writing A_3^{-1} in the form

$$\begin{bmatrix} (a^{-1} + b'n) & c'm \\ d'n & (h^{-1} + k'm) \end{bmatrix},$$

where

$$b' = \frac{h+km-a^{-1}\Delta}{\Delta n}, \quad c' = -\frac{c}{\Delta}$$

$$d' = -\frac{d}{\Delta}, \quad k' = \frac{a+bn-h^{-1}\Delta}{\Delta m}$$

$$\Delta = (a+bn)(h+km) - cdnm,$$

produces the inverse

$$B^{-1} = \begin{bmatrix} (a^{-1}I_n + b'J_{nn}) & c'J_{nm} \\ d'J_{mn} & (h^{-1}I_m + k'J_{mm}) \end{bmatrix}.$$

5. The next method we shall discuss is due to Newberry [41]. Consider a matrix of the form

$$Q = \begin{bmatrix} S & R \\ C & D \end{bmatrix}$$

where S is a scalar, R is a row-vector $[r_2, r_3, \ldots, r_n]$, C is a column-vector $[c_2, c_3, \ldots, c_n]^T$, and D is a diagonal matrix with diagonal elements $d_2, d_3, \ldots, d_n$. The inverse can be written in the form

$$Q^{-1} = \begin{bmatrix} S' & R' \\ C' & M' \end{bmatrix}$$

where each submatrix of Q^{-1} has the same form as the corresponding submatrix of Q except that M' is, in general, not diagonal. It can be shown that

$$S' = \left(S - \sum_{i=2}^{n} r_i c_i / d_i\right)^{-1}$$

and, for i,j = 2, 3, ..., n,

$$c_i' = -S'c_i/d_i$$

$$r_i' = -S'r_i/d_i$$

$$M_{ij}' = (\delta_{ij} - c_i r_j')/d_i \quad .$$

Let λ be an eigenvalue of Q, and let

$$x = \begin{bmatrix} 1 \\ x_2 \\ \vdots \\ x_n \end{bmatrix}$$

be an associated eigenvector. Then the equation $Qx = \lambda x$ leads to the following set of n equations:

$$S + \sum_{i=2}^{n} r_i x_i = \lambda$$

$$c_i + d_i x_i = \lambda x_i, \quad i = 2, 3, \ldots, n.$$

Eliminating the x_i yields

$$S + \sum_{i=2}^{n} r_i c_i/(\lambda - d_i) - \lambda = 0. \tag{1}$$

If we write

$$P(\lambda) = \prod_{i=2}^{n} (\lambda - d_i)$$

and

$$P_i(\lambda) = P(\lambda)/(\lambda - d_i), \quad i = 2, 3, \ldots, n,$$

then (1) can be written

$$(\lambda - S)P(\lambda) - \sum_{i=2}^{n} r_i c_i P_i(\lambda) = 0.$$

This is the characteristic equation of Q, and the following statements can be made concerning the eigenvalues:

(a) If all $r_i c_i > 0$ and all d_i are distinct, then all the eigenvalues are real and are separated by the d_i.

(b) If all d_i are equal to d, then d is an eigenvalue of multiplicity n - 2. The remaining two eigenvalues are the zeros of

$$\lambda^2 - (S+d)\lambda + Sd - \sum_{i=2}^{n} r_i c_i \tag{2}$$

and are real if, and only if,

$$(S-d)^2 + 4 \sum_{i=2}^{n} r_i c_i \geq 0.$$

(c) If all d_i are equal to d, the eigenvectors associated with the multiple eigenvalue d have zero as their first component and are orthogonal to the vector $[0, r_2, r_3, \ldots, r_n]$. If λ_p is a zero of (2), the eigenvector corresponding to λ_p is

$$\begin{bmatrix} \lambda_p - d \\ c_2 \\ c_3 \\ \vdots \\ c_n \end{bmatrix}.$$

6. Cline [13] also describes a general class of matrices with complex elements for which the inverse, eigenvalues, and eigenvectors are known. Let k be any real number such that $k \neq -1$. Let I be the identity matrix of order n, and let B be any matrix with complex elements having n columns. Suppose further that B has orthonormal rows. Then it follows that

$$(I+kB^H B)^{-1} = I - \frac{k}{k+1} B^H B.$$

Since

$$(I+kB^HB)B^H = (1+k)B^H$$

and

$$(I+kB^HB)(I-B^HB) = I - B^HB,$$

it follows that the columns of B^H provide an orthonormal set of eigenvectors of $(I+kB^HB)$ corresponding to the eigenvalue $\lambda = 1 + k$ and that the columns of $(I-B^HB)$ contain a linearly independent set of eigenvectors corresponding to the eigenvalue $\lambda = 1$. Now the rank of $(I+kB^HB)$ is equal to the rank of B^H plus the rank of $(I-B^HB)$. Thus B^H and any maximal linearly independent set of columns of $(I-B^HB)$ form a complete set of eigenvectors. It should also be pointed out that $(I+kB^HB)$ is Hermitian.

By taking B as the 1 x n matrix $[n^{-\frac{1}{2}}, n^{-\frac{1}{2}}, \ldots, n^{-\frac{1}{2}}]$ and $k = n/(d-1)$ where $d \neq 1$ and $d \neq -(n-1)$, we can obtain the test matrix of Pei [46]:

$$T = [t_{ij}]$$

where

$$t_{ij} = \begin{cases} d, & \text{if } i = j \\ 1, & \text{if } i \neq j. \end{cases}$$

We can write

$$\begin{aligned} T &= (d-1)I + nB^HB \\ &= (d-1)(I+kB^HB) \end{aligned}$$

and

$$\begin{aligned} T^{-1} &= \frac{1}{d-1}\left(I- \frac{k}{k+1} B^HB\right) \\ &= \frac{1}{d-1}\left(I- \frac{n}{n+d-1} B^HB\right). \end{aligned}$$

Thus, if $T^{-1} = [s_{ij}]$, we have

$$s_{ij} = \begin{cases} \dfrac{d+n-2}{d(d+n-2)-(n-1)}, & \text{if } i = j \\[2ex] \dfrac{-1}{d(d+n-2)-(n-1)}, & \text{if } i \neq j. \end{cases}$$

Furthermore, the eigenvalues of T are

$$\lambda = (d-1)(1+k)$$

$$= n + d - 1$$

of multiplicity one and $\lambda = d - 1$ of multiplicity n-1. Also,

$$\frac{1}{\sqrt{n}}\begin{bmatrix} 1 \\ 1 \\ 1 \\ 1 \\ \vdots \\ 1 \end{bmatrix}, \quad \frac{1}{\sqrt{2}}\begin{bmatrix} 1 \\ -1 \\ 0 \\ 0 \\ \vdots \\ 0 \end{bmatrix}, \quad \frac{1}{\sqrt{6}}\begin{bmatrix} 1 \\ 1 \\ -2 \\ 0 \\ \vdots \\ 0 \end{bmatrix}, \quad \ldots, \quad \frac{1}{\sqrt{n(n-1)}}\begin{bmatrix} 1 \\ 1 \\ 1 \\ \vdots \\ 1 \\ -(n-1) \end{bmatrix}$$

form an orthonormal set of eigenvectors of T, where the first corresponds to the eigenvalue $\lambda = n + d - 1$.

7. Ortega [44] describes a valuable method using similarity transformations to generate test matrices. Let $C = I + uv^H$ where u and v are n x 1 column-vectors. Then

$$C^{-1} = I - (1+v^H u)^{-1} uv^H.$$

It can be shown that any vector orthogonal to v is an eigenvector of C corresponding to the eigenvalue $\lambda = 1$ and that u is an eigenvector of C corresponding to the eigenvalue $\lambda = 1 + v^H u$. Since the eigenvalue $\lambda = 1$ has multiplicity n-1, the matrices C have limited use in testing eigenvalue routines. They are quite useful, however, in testing inversion procedures.

Now let $\alpha = (1+v^H u)^{-1}$, and let R be any n x n matrix. Then the similarity transformation $A = CRC^{-1}$ becomes

$$A = (I+uv^H)R(I+uv^H)^{-1}$$

$$= R + uv^H R - \alpha Ruv^H - \alpha(v^H Ru)uv^H.$$

The inverse is given by

$$A^{-1} = CR^{-1}C^{-1}$$

$$= R^{-1} + uv^H R^{-1} - \alpha R^{-1}uv^H - \alpha(v^H R^{-1}u)uv^H.$$

Proper selection of u, v, and R will insure that A and A^{-1} can be generated exactly in the computer. To illustrate the possibilities, we present the following examples. For simplicity we consider only real u, v, and R unless stated otherwise.

Real symmetric matrices can be generated by letting

$$u = -2v, \quad \sum_{i=1}^{n} v_i = 1$$

$$R = D = \text{diag}[d_1, d_2, \ldots, d_n].$$

Then $(I-2vv^T)$ is orthogonal, and

$$A = (I-2vv^T)D(I-2vv^T)$$

$$= D - 2vv^T D - 2Dvv^T + 4(v^T Dv)vv^T.$$

The matrix A is symmetric, has eigenvalues $d_1, d_2, \ldots, d_n$, and eigenvectors which are the columns of $(I-2vv^T)$. In particular, if $A = [a_{ij}]$ and

$$v^T = [n^{-\frac{1}{2}}, n^{-\frac{1}{2}}, \ldots, n^{-\frac{1}{2}}],$$

then

$$a_{ij} = n^{-1}(nd_i\delta_{ij} - 2d_i - 2d_j + 2r)$$

where

$$r = 2n^{-1}\sum_{k=1}^{n} d_k.$$

If we let $R = \text{diag}[R_1, R_2, \ldots, R_p]$ be a block-diagonal matrix such that R_i is a complex Hermitian matrix, then

$$A = (I-2vv^T)R(I-2vv^T)$$

will also be Hermitian.

To generate nonsymmetric real matrices, we have a much wider choice for u and v, although the restriction $u^Tv = 0$ affords some simplification. For example, if n is even, say $n = 2k$, $u^T = c[1,1,\ldots,1]$, $v^T = [1,1,\ldots,1,-1,-1,\ldots,-1]$ with k components of 1 and k components of -1, and $\sigma = v^TDu$, then

$$A = D + uv^TD - Duv^T - \sigma uv^T.$$

It can be shown that, if $A = [a_{ij}]$,

$$a_{ij} = \begin{cases} d_i\delta_{ij} - c(d_i - d_j + \sigma), & i \le j \le k \\ \\ d_i\delta_{ij} + c(d_i - d_j + \sigma), & k+1 \le j \le n. \end{cases}$$

The matrix A has real eigenvalues $d_1, d_2, \ldots, d_n$, right eigenvectors which are the columns of $(I+uv^T)$, and left eigenvectors which are the rows of $(I-uv^T)$. It is easy to generate A exactly since only additions of the d_i and multiplications by c are involved; if $c = 1$, only additions are required. The parameter c provides some control over the condition of the problem since the n condition numbers of A are

$$\{[(1+c)^2 + (n-1)c^2][(1-c)^2 + (n-1)c^2]\}^{-\frac{1}{2}} .$$

Another choice for u and v is $u^T = [1, 2, \ldots, k, 1, 2, \ldots, k]$ and $v^T = [1, 2, \ldots, k, -1, -2, \ldots, -k]$. The relation $u^T v = 0$ is maintained, and the n condition numbers are

$$s_m = s_{k+m} = \{[1+2m^2(\beta+1)][1+2m^2(\beta-1)]\}^{-\frac{1}{2}},$$

where $m = 1, 2, \ldots, k$, and $\beta = \sum_{i=1}^{k} i^2$.

Now if we let $R = \mathrm{diag}[R_1, R_2, \ldots, R_p]$ be a block-diagonal matrix, we can obtain a real matrix A which has complex eigenvalues. For example, the R_i may be 2 x 2 real matrices which have known complex eigenvalues.

8. Next we consider a family of matrices, called circulants, which are of the form

$$C = \begin{bmatrix} c_0 & c_1 & c_2 & \cdots & c_{n-1} \\ c_{n-1} & c_0 & c_1 & \cdots & c_{n-2} \\ c_{n-2} & c_{n-1} & c_0 & \cdots & c_{n-3} \\ \cdots & \cdots & \cdots & \cdots & \cdots \\ c_1 & c_2 & c_3 & \cdots & c_0 \end{bmatrix} .$$

Let $r_k = \exp \frac{2k\pi i}{n}$, $k = 1, 2, \ldots, n$, be the solutions of the equation $r^n = 1$. Then it can be shown [2, pp. 234-235] that, for $k = 1, 2, \ldots, n$,

$$y_k = c_0 + c_1 r_k + c_2 r_k^2 + \ldots + c_{n-1} r_k^{n-1}$$

is an eigenvalue of C with associated right eigenvector

$$x^{(k)} = \begin{bmatrix} 1 \\ r_k \\ r_k^2 \\ \vdots \\ r_k^{n-1} \end{bmatrix}$$

and left eigenvector

$$y^{(k)} = [r_k^{n-1}, r_k^{n-2}, \ldots, r_k, 1].$$

Observe that since the r_k are all distinct, C has n distinct right eigenvectors and n distinct left eigenvectors. Note also that if we choose $c_i = c_{n-i}$, the matrix C is symmetric. Circulants can be generalized [2, p. 235] by using equations of the form

$$r^n = b_1 r^{n-1} + b_2 r^{n-2} + \ldots + b_n.$$

9. Brenner [6] has defined a set of matrices related to the Mahler matrices [35] for which the determinant, eigenvalues, eigenvectors, and elementary divisors are known. Let k and n be positive integers such that $(k,n) = 1$, $k > 1$, $n > 1$, and let m be a positive integer with $(m,n) = 1$, $m \equiv 1 \pmod{k}$. Set $s = \exp \frac{2\pi i}{k}$, and let Q be the n x n matrix defined by

$$Q = \begin{bmatrix} 0 & 1 & 0 & \cdots & 0 & 0 \\ 0 & 0 & 1 & \cdots & 0 & 0 \\ & & \cdots\cdots & \cdots\cdots & & \\ 0 & 0 & 0 & \cdots & 0 & 1 \\ s & 0 & 0 & \cdots & 0 & 0 \end{bmatrix}.$$

Define v to be the 1 x n row-vector

$$\frac{1}{1-s}[1, 1, \ldots, 1].$$

Finally let A(m) denote the n x n matrix defined by

$$A(m) = \begin{bmatrix} v(I-Q^{m}) \\ v(Q^{m}-Q^{2m}) \\ \cdots\cdots\cdots\cdots \\ v(Q^{(n-1)m}-Q^{nm}) \end{bmatrix}$$

where the r-th row of A(m) is $v(Q^{(r-1)m}-Q^{rm})$, $r = 1, 2, \ldots, n$. It should be pointed out that if $k = 2$, the elements of A(m) are 1, 0, -1.

It can be shown that there exists a non-singular matrix M such that $M^{-1}A(m)M$ is a block-diagonal matrix. To obtain M and the block-diagonal form, we need the following definitions. Let

$$\omega(t) = \exp[2\pi i(kt-k+1)/kn].$$

Define the function gt mod n of t mod n by

$$gt \equiv mt + (k-1)[(m-1)/k] \pmod{n}.$$

It follows that, if r is a positive integer,

$$g^{r}t \equiv m^{r}t + (k-1)[(m^{r}-1)/k] \pmod{n}.$$

Next let w(t,A(m)) be the function defined by

$$w(t,A(m)) = \frac{1-\omega^{m}(t)}{1-\omega(t)}.$$

Now define an equivalence relation "~" on the residue classes mod n by $t_1 \sim t_2$ if, and only if, $t_1 \equiv g^r t_2$ for some positive integer r. Let $t_1, t_2, \ldots, t_p$ be representative members of the equivalence classes of the relation "~," and let n_j, j = 1, 2, ..., p, be the number of elements in the j-th equivalence class. Note that the j-th class consists of the objects $gt_j, g^2 t_j, \ldots, g^{n_j} t_j \equiv t_j$.

Let x(t) be the n x 1 column-vector given by

$$x(t) = \begin{bmatrix} 1 \\ \omega(t) \\ \omega^2(t) \\ \vdots \\ \omega^{n-1}(t) \end{bmatrix}, \quad t = 1, 2, \ldots, n.$$

Define the matrix M by

$$M = [X(t_1), X(t_2), \ldots, X(t_p)]$$

where

$$X(t_j) = [x(gt_j), x(g^2 t_j), \ldots, x(g^{n_j} t_j)], \quad j = 1, 2, \ldots, p.$$

Then $C = M^{-1}A(m)M$ is the block-diagonal matrix

$$\text{diag}[W(t_1,A(m)), W(t_2,A(m)), \ldots, W(t_p,A(m))]$$

where, for j = 1, 2, ..., p, $W(t_j,A(m))$ is the n_j x n_j matrix defined by

$$\begin{bmatrix} 0 & 0 & \cdots & 0 & w(g^{n_j}t_j,A(m)) \\ w(gt_j,A(m)) & & \cdots & 0 & 0 \\ 0 & w(g^2t_j,A(m)) & \cdots & 0 & 0 \\ \cdots & \cdots & \cdots & \cdots & \cdots \\ 0 & 0 & \cdots & w(g^{n_j-1}t_j,A(m)) & 0 \end{bmatrix}.$$

To obtain the eigenvalues and eigenvectors, we make use of the fact that the n x n matrix

$$D = \begin{bmatrix} 0 & 0 & \cdots & 0 & a_n \\ a_1 & 0 & \cdots & 0 & 0 \\ 0 & a_2 & \cdots & 0 & 0 \\ \cdots & \cdots & \cdots & \cdots & \cdots \\ 0 & 0 & \cdots & a_{n-1} & 0 \end{bmatrix}$$

where $a_n a_{n-1} a_{n-2} \cdots a_1 = 1$, has as eigenvalues the n n-th roots of unity and that the eigenvector of D associated with the eigenvalue λ is

$$\begin{bmatrix} \lambda^n \\ a_1\lambda^{n-1} \\ a_2a_1\lambda^{n-2} \\ \vdots \\ a_{n-1}a_{n-2}\cdots a_1\lambda \end{bmatrix}.$$

Since the matrices $W(t_j,A(m))$ are all of the same form as D, we can obtain the eigenvalues and eigenvectors of A(m) from the block-diagonal form C by applying Theorems 6 and 7.

As an example, let n = 5, m = 4, k = 3. Then

$$A(4) = \begin{bmatrix} 1 & 1 & 1 & 1 & 0 \\ s & s & s & 0 & 1 \\ s^2 & s^2 & 0 & s & s \\ 1 & 0 & s^2 & s^2 & s^2 \\ 0 & 1 & 1 & 1 & 1 \end{bmatrix}$$

where $s = \exp \frac{2\pi i}{3}$. The relation "$\sim$" has three equivalence classes, {0,2}, {1}, {3,4}. Hence $n_1 = n_3 = 2$ and $n_2 = 1$. Thus the eigenvalues of A(4) are the square roots of unity (each twice) and unity: 1, 1, 1, -1, -1.

We close this chapter by describing the Vandermonde matrices [34] for which there is an explicit representation of the inverse. Let

$$V(x_1, x_2, \ldots, x_n) = \begin{bmatrix} 1 & 1 & \cdots & 1 \\ x_1 & x_2 & \cdots & x_n \\ x_1^2 & x_2^2 & \cdots & x_n^2 \\ \cdots & \cdots & \cdots & \cdots \\ x_1^{n-1} & x_2^{n-1} & \cdots & x_n^{n-1} \end{bmatrix}$$

where the x_i are distinct and non-zero. Then if we let $V^{-1}(x_1, x_2, \ldots, x_n) = [v_{ij}]$, we have

$$v_{ij} = x_i b_{ij}$$

where

$$b_{ij} = \frac{(-1)^{n-j} \sigma^{i}_{n-j,n-1}}{\prod_{\substack{k=0 \\ k \neq i}}^{n} (x_i - x_k)}$$

and $\sigma^{i}_{m,n-1}$ is the sum of all products of m of the numbers $x_1, x_2, \ldots, x_{i-1}, x_{i+1}, \ldots, x_n$ without permutations or repetitions ($\sigma^{i}_{0,n-1} = 1$, $x_0 = 0$). For example,

$$V(1,2,3,4) = \begin{bmatrix} 1 & 1 & 1 & 1 \\ 1 & 2 & 3 & 4 \\ 1 & 4 & 9 & 16 \\ 1 & 8 & 27 & 64 \end{bmatrix}$$

has the inverse

$$V^{-1}(1,2,3,4) = \begin{bmatrix} 4 & -\frac{13}{3} & \frac{3}{2} & -\frac{1}{6} \\ -6 & \frac{19}{2} & -4 & \frac{1}{2} \\ 4 & -7 & \frac{7}{2} & -\frac{1}{2} \\ -1 & \frac{11}{6} & -1 & \frac{1}{6} \end{bmatrix}.$$

10. Forsythe [78] points out that Varah [79] "also has a program for generating an arbitrary matrix example, starting from the Jordan form, and subject to the round-off in inverting the matrix of eigenvectors (+ principal vectors), and in multiplying matrices. Where he starts with integers and enough precision, and where the determinant is a small integer, you can see that there will be no round-off error at all." See examples 5.25, 5.26, and 5.27.

CHAPTER III

TEST MATRICES: INVERSES, SYSTEMS OF LINEAR EQUATIONS, AND DETERMINANTS

Example 3.1

$$A = \begin{bmatrix} 33 & 16 & 72 \\ -24 & -10 & -57 \\ -8 & -4 & -17 \end{bmatrix} \qquad A^{-1} = \frac{1}{6}\begin{bmatrix} -58 & -16 & -192 \\ 48 & 15 & 153 \\ 16 & 4 & 54 \end{bmatrix}$$

$$\text{If } b = \begin{bmatrix} -359 \\ 281 \\ 85 \end{bmatrix}, \text{ then } x = A^{-1}b = \begin{bmatrix} -1 \\ 2 \\ 5 \end{bmatrix}.$$

Reference: [29, pp. 120-122].

Example 3.2

$$A = \begin{bmatrix} 1 & -2 & 3 & 1 \\ -2 & 1 & -2 & -1 \\ 3 & -2 & 1 & 5 \\ 1 & -1 & 5 & 3 \end{bmatrix} \qquad A^{-1} = \frac{1}{52} \cdot \begin{bmatrix} -15 & -38 & -1 & -6 \\ -38 & -20 & -6 & 16 \\ -1 & -6 & -7 & 10 \\ -6 & 16 & 10 & 8 \end{bmatrix}$$

$$\text{If } b = \begin{bmatrix} 3 \\ -4 \\ 7 \\ 8 \end{bmatrix}, \text{ then } x = A^{-1}b = \begin{bmatrix} 1 \\ 1 \\ 1 \\ 1 \end{bmatrix}.$$

Reference: [4, p. 100].

Example 3.3

$$A = \begin{bmatrix} 1 & 1+2i & 2+10i \\ 1+i & 3i & -5+14i \\ 1+i & 5i & -8+20i \end{bmatrix}$$

$$A^{-1} = \begin{bmatrix} 10+i & -2+6i & -3-2i \\ 9-3i & 8i & -3-2i \\ -2+2i & -1-2i & 1 \end{bmatrix}$$

Reference: [60, p. 136].

Example 3.4

$$A = \begin{bmatrix} 1 & 0 & 0 & 0 & 0 & 1 \\ 1 & 1 & 0 & 0 & 0 & -1 \\ -1 & 1 & 1 & 0 & 0 & 1 \\ 1 & -1 & 1 & 1 & 0 & -1 \\ -1 & 1 & -1 & 1 & 1 & 1 \\ 1 & -1 & 1 & -1 & 1 & -1 \end{bmatrix}$$

$$A^{-1} = \begin{bmatrix} 2^{-1} & 2^{-2} & -2^{-3} & 2^{-4} & -2^{-5} & 2^{-5} \\ 0 & 2^{-1} & 2^{-2} & -2^{-3} & 2^{-4} & -2^{-4} \\ 0 & 0 & 2^{-1} & 2^{-2} & -2^{-3} & 2^{-3} \\ 0 & 0 & 0 & 2^{-1} & 2^{-2} & -2^{-2} \\ 0 & 0 & 0 & 0 & 2^{-1} & 2^{-1} \\ 2^{-1} & -2^{-2} & 2^{-3} & -2^{-4} & 2^{-5} & -2^{-5} \end{bmatrix}$$

Reference: [64].

Example 3.5

$$A = \begin{bmatrix} 5 & 7 & 6 & 5 \\ 7 & 10 & 8 & 7 \\ 6 & 8 & 10 & 9 \\ 5 & 7 & 9 & 10 \end{bmatrix}, \quad A^{-1} = \begin{bmatrix} 68 & -41 & -17 & 10 \\ -41 & 25 & 10 & -6 \\ -17 & 10 & 5 & -3 \\ 10 & -6 & -3 & 2 \end{bmatrix}$$

$$\text{If } b = \begin{bmatrix} 23 \\ 32 \\ 33 \\ 31 \end{bmatrix}, \quad \text{then } x = \begin{bmatrix} 1 \\ 1 \\ 1 \\ 1 \end{bmatrix}.$$

Condition numbers:

$$M(A) = 2720$$
$$N(A) = 752$$
$$P(A) = 2984$$

Reference: [23], [36].

Example 3.6 (See also Example 3.23.)

Let $A = [a_{ij}]$ be the n x n matrix defined by

$$a_{ij} = n - |i - j|.$$

Then $A^{-1} = [b_{ij}]$ is given by

$$b_{ij} = \begin{cases} \frac{n+2}{2n+2}, & \text{if } i = j = 1 \text{ or } i = j = n \\ 1, & \text{if } i = j \text{ and } 1 < i < n \\ -\frac{1}{2}, & \text{if } |i - j| = 1 \text{ and } n \neq 2 \\ -\frac{1}{3}, & \text{if } |i - j| = 1 \text{ and } n = 2 \\ \frac{1}{2n+2}, & \text{if } |i - j| = n - 1 \neq 1 \\ 0, & \text{if } 1 < |i - j| < n - 1 \end{cases}$$

For example, when n = 4,

$$A = \begin{bmatrix} 4 & 3 & 2 & 1 \\ 3 & 4 & 3 & 2 \\ 2 & 3 & 4 & 3 \\ 1 & 2 & 3 & 4 \end{bmatrix}$$

$$A^{-1} = \begin{bmatrix} \frac{3}{5} & -\frac{1}{2} & 0 & \frac{1}{10} \\ -\frac{1}{2} & 1 & -\frac{1}{2} & 0 \\ 0 & -\frac{1}{2} & 1 & -\frac{1}{2} \\ \frac{1}{10} & 0 & -\frac{1}{2} & \frac{3}{5} \end{bmatrix}$$

Reference: [31], [60, p. 137].

Example 3.7 (Pascal's Matrix)

Let $A = [a_{ij}]$ be the n x n matrix defined by

$$a_{1j} = a_{j1} = k \neq 0, \qquad j = 1, 2, \ldots, n,$$

$$a_{ij} = a_{i-1,j} + a_{i,j-1}, \quad i,j = 2, 3, \ldots, n.$$

Equivalently, we have, for i,j = 1, 2, ..., n,

$$a_{ij} = k \frac{(i+j-2)!}{(i-1)!(j-1)!} .$$

This is called Pascal's matrix because the coefficients of k are obtained from the Pascal triangle of binomial coefficients.

If k is the reciprocal of an integer, the elements of A^{-1} are integers. In addition, $\det(A) = k^n$.

For example, if $n = 4$ and $k = \frac{1}{7}$,

$$A = \begin{bmatrix} \frac{1}{7} & \frac{1}{7} & \frac{1}{7} & \frac{1}{7} \\ \frac{1}{7} & \frac{2}{7} & \frac{3}{7} & \frac{4}{7} \\ \frac{1}{7} & \frac{3}{7} & \frac{6}{7} & \frac{10}{7} \\ \frac{1}{7} & \frac{4}{7} & \frac{10}{7} & \frac{20}{7} \end{bmatrix}$$

$$A^{-1} = \begin{bmatrix} 28 & -42 & 28 & -7 \\ -42 & 98 & -77 & 21 \\ 28 & -77 & 70 & -21 \\ -7 & 21 & -21 & 7 \end{bmatrix}$$

$$\det(A) = \frac{1}{7^4} = \frac{1}{2401}$$

Reference: [11].

<u>Example 3.8</u> (The finite segments of the (infinite) Hilbert Matrix. See example 3.26 for a generalization.)

Let $A_n = \left[a_{ij}^{(n)}\right]$ be the n x n matrix defined by

$$a_{ij}^{(n)} = \frac{1}{i+j-1}, \qquad i,j = 1, 2, \ldots, n.$$

$$A_n = \begin{bmatrix} 1 & \frac{1}{2} & \frac{1}{3} & \cdots & \frac{1}{n} \\ \frac{1}{2} & \frac{1}{3} & \frac{1}{4} & \cdots & \frac{1}{n+1} \\ \frac{1}{3} & \frac{1}{4} & \frac{1}{5} & \cdots & \frac{1}{n+2} \\ \cdots & \cdots & \cdots & \cdots & \cdots \\ \frac{1}{n} & \frac{1}{n+1} & \frac{1}{n+2} & \cdots & \frac{1}{2n-1} \end{bmatrix}$$

If $A_n^{-1} = \left[b_{ij}^{(n)}\right]$, then

$$b_{ij}^{(n)} = \frac{(-1)^{i+j}(n+i-1)!(n+j-1)!}{(i+j-1)[(i-1)!(j-1)!]^2(n-i)!(n-j)!}$$

Alternatively, if $b_{11}^{(1)} = 1$, then for $n \geq 1$,

$$b_{ij}^{(n+1)} = \frac{(n+i)(n+j)}{(n+1-i)(n+1-j)} b_{ij}^{(n)}, \qquad i,j = 1, 2, \ldots, n,$$

$$b_{n+1,j}^{(n+1)} = b_{j,n+1}^{(n+1)}$$

$$= \frac{(-1)^{n+j-1}(2n+1)!(n+j)!}{(n+j)[n!(j-1)!]^2(n+1-j)!}, \qquad j = 1, 2, \ldots, n+1.$$

Since A_n^{-1} is symmetric, we do not exhibit the elements above the main diagonal in the following matrices:

$$A_2^{-1} = \begin{bmatrix} 4 & \\ -6 & 12 \end{bmatrix} \qquad A_3^{-1} = \begin{bmatrix} 9 & & \\ -36 & 192 & \\ 30 & -180 & 180 \end{bmatrix}$$

$$A_4^{-1} = \begin{bmatrix} 16 & & & \\ -120 & 1200 & & \\ 240 & -2700 & 6480 & \\ -140 & 1680 & -4200 & 2800 \end{bmatrix}$$

$$A_5^{-1} = \begin{bmatrix} 25 & & & & \\ -300 & 4800 & & & \\ 1050 & -18900 & 79380 & & \\ -1400 & 26880 & -117600 & 179200 & \\ 630 & -12600 & 56700 & -88200 & 44100 \end{bmatrix}$$

$$A_6^{-1} = \begin{bmatrix} 36 & & & & & \\ -630 & 14700 & & & & \\ 3360 & -88200 & 564480 & & & \\ -7560 & 211680 & -1411200 & 3628800 & & \\ 7560 & -220500 & 1512000 & -3969000 & 4410000 & \\ -2772 & 83160 & -582120 & 1552320 & -1746360 & 698544 \end{bmatrix}$$

$$A_7^{-1} = \begin{bmatrix} 49 & & & & & & \\ -1176 & 37632 & & & & & \\ 8820 & -317520 & 2857680 & & & & \\ -29400 & 1128960 & -10584000 & 40320000 & & & \\ 48510 & -1940400 & 18711000 & -72765000 & 133402500 & & \\ -38808 & 1596672 & -15717240 & 62092800 & -115259760 & 100590336 & \\ 12012 & -504504 & 5045040 & -20180160 & 37837800 & -33297264 & 11099088 \end{bmatrix}$$

$$A_8^{-1} = \begin{bmatrix} 64 & & & & & & & \\ -2016 & 84672 & & & & & & \\ 20160 & -952560 & 11430720 & & & & & \\ -92400 & 4656960 & -58212000 & 304920000 & & & & \\ 221760 & -11642400 & 149688000 & -800415000 & 2134440000 & & & \\ -288288 & 15567552 & -204324120 & 1109908800 & -2996753760 & 4249941696 & & \\ 192192 & -10594584 & 141261120 & -776936160 & 2118916800 & -3030051024 & 2175421248 & \\ -51480 & 2882880 & -38918880 & 216216000 & -594594000 & 856215360 & -618377760 & 176679360 \end{bmatrix}$$

$$A_9^{-1} = \begin{bmatrix} 81 & & & & & & & & \\ -3240 & 172800 & & & & & & & \\ 41580 & -2494800 & 38419920 & & & & & & \\ -249480 & 15966720 & -256132800 & 1756339200 & & & & & \\ 810810 & -54054000 & 891891000 & -6243237000 & 22545022500 & & & & \\ -1513512 & 103783680 & -1748106360 & 12430978560 & -45450765360 & 92554285824 & & & \\ 1621620 & -113513400 & 1942340400 & -13984850880 & 51648597000 & -106051785840 & 122367445200 & & \\ -926640 & 65894400 & -1141620480 & 8302694400 & -30918888000 & 63930746880 & -74205331200 & 45229916160 & \\ 218790 & -15752880 & 275675400 & -2021619600 & 7581073500 & -15768632880 & 18396738360 & -11263309200 & 2815827300 \end{bmatrix}$$

$$A_{10}^{-1} = \begin{bmatrix}
100 & & & & \\
-4950 & 326700 & & & \\
79200 & -5880600 & 112907520 & & \\
-600600 & 47567520 & -951350400 & 8245036800 & \\
2522520 & -208107900 & 4281076800 & -37875637800 & \cdots \\
-6306300 & 535134600 & -11237826600 & 101001700800 & \\
9609600 & -832431600 & 17758540800 & -161602721280 & \\
-8751600 & 770140800 & -16635041280 & 152907955200 & \\
4375800 & -389883780 & 8506555200 & -78843164400 & \\
-923780 & 83140200 & -1829084400 & 17071454400 &
\end{bmatrix}$$

$$\begin{matrix}
 & 176752976400 & & & \\
 & -477233036280 & 1301544644400 & & \\
\cdots & 771285715200 & -2121035716800 & 3480673996800 & \cdots \\
 & -735869534400 & 2037792556800 & -3363975014400 & \\
 & 382086104400 & -1064382719400 & 1766086882560 & \\
 & -83223340200 & 233025352560 & -388375587600 &
\end{matrix}$$

$$\begin{bmatrix}
 & 3267861442560 & & \\
\cdots & -1723286307600 & 912328045200 & \\
 & 380449555200 & -202113826200 & 44914183600
\end{bmatrix}$$

Condition numbers:

$$P(A_n) \doteq e^{3.5\, n}$$

$$M(A_n) \sim ke^{3.525\, n}$$

where k is a constant.

Determinant:

n	$\det(A_n)$
2	8.33333 33333 33333 33333 33333 (-2)
3	4.62962 96296 29629 62962 96296 (-4)
4	1.65343 91534 39153 43915 34392 (-7)
5	3.74929 51325 15087 16361 32407 (-12)
6	5.36729 98873 58687 73278 88304 (-18)
7	4.83580 26239 26116 93211 98556 (-25)
8	2.73705 01137 91513 01664 20433 (-33)
9	9.72023 43119 24999 86288 94723 (-43)
10	2.16417 92264 31491 86906 05950 (-53)

Reference: [20], [43], [50], [55].

Example 3.9

Let $A_n = [a_{ij}]$ be the n x n matrix given by

$$a_{1j} = 1, \qquad j = 1, 2, \ldots, n,$$

$$a_{ij} = (i+j-1)^{-1}, \qquad i = 2, 3, \ldots, n, \quad j = 1, 2, \ldots, n.$$

$$A_n = \begin{bmatrix} 1 & 1 & 1 & \cdots & 1 \\ \frac{1}{2} & \frac{1}{3} & \frac{1}{4} & \cdots & \frac{1}{n+1} \\ \frac{1}{3} & \frac{1}{4} & \frac{1}{5} & \cdots & \frac{1}{n+2} \\ & \cdots & \cdots & \cdots & \\ \frac{1}{n} & \frac{1}{n+1} & \frac{1}{n+2} & \cdots & \frac{1}{2n-1} \end{bmatrix}$$

If $A_n^{-1} = [b_{ij}]$, then

$$b_{i1} = (-1)^{n-i}\binom{n+i-1}{i-1}\binom{n}{i}, \qquad i = 1, 2, \ldots, n,$$

$$b_{i,j+1} = (-1)^{i-j}\binom{i+j}{j}\binom{i+j-1}{j-1}\binom{n+i-1}{i+j}\binom{n+j}{i+j}\, i,$$

for $i = 1, 2, \ldots, n$, $j = 1, 2, \ldots, n-1$.

Furthermore,

$$\det(A_n) = (-1)^{n-1}\delta_n^{-1},$$

where

$$\delta_{n+1} = \binom{2n}{n-1}\binom{2n}{n}(2n+1)\delta_n \quad \text{and} \quad \delta_1 = 1.$$

The condition numbers of A_n are given by

$$P(A_n) \doteq C_1\, 2^{5n} \log n$$

$$M(A_n) \doteq C_2\, n\, 2^{5n}$$

where $C_1 \doteq 8 \times 10^{-3}$ and $C_2 \doteq 4 \times 10^{-3}$. The condition numbers of A_n, $n = 2, 3, \ldots, 10$, rounded to 5 significant digits, are given in the following table:

n	$M(A_n)$	$P(A_n)$
2	12	12.587
3	540	354.51
4	17280	13090
5	67200×10^1	45057×10^1
6	23814×10^3	15259×10^3
7	80681×10^4	51270×10^4
8	28333×10^6	17164×10^6
9	95447×10^7	57364×10^7
10	33640×10^9	19158×10^9

When $n = 6$, the inverse is given by

$$A_6^{-1} = \begin{bmatrix} -6 & 630 & -6720 & 22680 & -30240 & 13860 \\ 105 & -7350 & 88200 & -317520 & 441000 & -207900 \\ -560 & 29400 & -376320 & 1411200 & -2016000 & 970200 \\ 1260 & -52920 & 705600 & -2721600 & 3969000 & -1940400 \\ -1260 & 44100 & -604800 & 2381400 & -3528000 & 1746360 \\ 462 & -13860 & 194040 & -776160 & 1164240 & -582120 \end{bmatrix}$$

Reference: [33].

Example 3.10

Let $A = [a_{ij}]$ be the n x n matrix defined by

$$a_{ij} = \begin{cases} \frac{i}{j}, & \text{if } i \leq j, \\ \frac{j}{i}, & \text{if } i > j. \end{cases}$$

If $A^{-1} = [b_{ij}]$, then

$$b_{ij} = \begin{cases} \frac{4i^3}{4i^2-1}, & \text{if } i = j \text{ and } i < n \\ \frac{n^2}{2n-1}, & \text{if } i = j = n \\ -\frac{i(i+1)}{2i+1}, & \text{if } j = i + 1 \\ -\frac{j(j+1)}{2j+1}, & \text{if } i = j + 1 \\ 0, & \text{if } |i-j| > 1. \end{cases}$$

For example, when n = 3,

$$A = \begin{bmatrix} 1 & \frac{1}{2} & \frac{1}{3} \\ \frac{1}{2} & 1 & \frac{2}{3} \\ \frac{1}{3} & \frac{2}{3} & 1 \end{bmatrix} \qquad A^{-1} = \begin{bmatrix} \frac{4}{3} & -\frac{2}{3} & 0 \\ -\frac{2}{3} & \frac{32}{15} & -\frac{6}{5} \\ 0 & -\frac{6}{5} & \frac{9}{5} \end{bmatrix}$$

Reference: [43], [60, pp. 138-139].

Example 3.11

Let $A = [a_{ij}]$ be the n x n matrix defined by

$$a_{ij} = \left(\frac{2}{n+1}\right)^{\frac{1}{2}} \sin\left(\frac{ij\pi}{n+1}\right) .$$

Then A is orthogonal, and $A^{-1} = A$.

Reference: [43].

Example 3.12

Let $A = [a_{ij}]$ be the n x n matrix defined by

$$a_{ij} = a_{ji} = n + 1 - i, \quad \text{if} \quad i \geq j.$$

$$A = \begin{bmatrix} n & n-1 & n-2 & \cdots & 2 & 1 \\ n-1 & n-1 & n-2 & \cdots & 2 & 1 \\ n-2 & n-2 & n-2 & \cdots & 2 & 1 \\ & \cdots & \cdots & \cdots & \cdots & \\ 2 & 2 & 2 & \cdots & 2 & 1 \\ 1 & 1 & 1 & \cdots & 1 & 1 \end{bmatrix}$$

$$A^{-1} = \begin{bmatrix} 1 & -1 & & & & \\ -1 & 2 & -1 & & & \\ & -1 & 2 & -1 & & \\ & \cdots & \cdots & \cdots & \cdots & \\ & & & -1 & 2 & -1 \\ & & & & -1 & 2 \end{bmatrix}, \quad n \times n.$$

Reference: [24].

Example 3.13

Let $A = [a_{ij}]$ be the n x n matrix defined by

$$a_{ij} = \max(i,j), \quad i,j = 1, 2, \ldots, n.$$

$$A = \begin{bmatrix} 1 & 2 & 3 & \cdots & n-1 & n \\ 2 & 2 & 3 & \cdots & n-1 & n \\ 3 & 3 & 3 & \cdots & n-1 & n \\ \cdots & \cdots & \cdots & \cdots & \cdots & \cdots \\ n-1 & n-1 & n-1 & \cdots & n-1 & n \\ n & n & n & \cdots & n & n \end{bmatrix}$$

$$A^{-1} = \begin{bmatrix} -1 & 1 & & & & \\ 1 & -2 & 1 & & & \\ & 1 & -2 & 1 & & \\ & \cdots & \cdots & \cdots & \cdots & \\ & & & 1 & -2 & 1 \\ & & & & 1 & -\frac{n-1}{n} \end{bmatrix}, \ n \times n.$$

Reference: [5].

Example 3.14 (Hadamard Matrices)

Let p be a prime, $p \geq 3$, and let $n = p - 1$. For an arbitrary integer k, define

$$\left(\frac{k}{p}\right) = \begin{cases} 0, & \text{if p divides k,} \\ 1, & \text{if k is congruent to a square mod p,} \\ -1, & \text{otherwise.} \end{cases}$$

Define $A = [a_{ij}]$ to be the n x n matrix such that

$$a_{ij} = \left(\frac{i+j}{p}\right), \quad i,j = 1, 2, \ldots, n.$$

If $A^{-1} = [b_{ij}]$, then

$$b_{ij} = \frac{1}{p}\left[\left(\frac{i+j}{p}\right) - \left(\frac{i}{p}\right) - \left(\frac{j}{p}\right)\right], \qquad i,j = 1, 2, \ldots, n.$$

For example, when $p = 5$,

$$A = \begin{bmatrix} -1 & -1 & 1 & 0 \\ -1 & 1 & 0 & 1 \\ 1 & 0 & 1 & -1 \\ 0 & 1 & -1 & -1 \end{bmatrix} \qquad A^{-1} = \frac{1}{5}\begin{bmatrix} -3 & -1 & 1 & -2 \\ -1 & 3 & 2 & 1 \\ 1 & 2 & 3 & -1 \\ -2 & 1 & -1 & -3 \end{bmatrix}$$

Condition number:

$$P(A) = \sqrt{p}$$

Reference: [43].

Example 3.15

Let B be the n x n matrix (row elements are binomial coefficients except for sign)

$$B = \begin{bmatrix} 1 & 0 & 0 & 0 & 0 & \cdots \\ 1 & -1 & 0 & 0 & 0 & \cdots \\ 1 & -2 & 1 & 0 & 0 & \cdots \\ 1 & -3 & 3 & -1 & 0 & \cdots \\ 1 & -4 & 6 & -4 & 1 & \cdots \\ \cdots & \cdots & \cdots & \cdots & \cdots & \end{bmatrix}.$$

Then $B^{-1} = B$. Furthermore, if $A = B^TB$, then $a_{ij} = \binom{i+j}{j}$, $i,j = 0,1,2,\ldots,n-1$, and $A^{-1} = BB^T$. The eigenvalues of B are all of modulus 1, and the eigenvalues of A occur in reciprocal pairs.

Condition number:

$$P(A) \doteq \exp(4n \log 2)$$

Reference: [42, pp. 240-241], [60, p. 140].

Example 3.16

$$A_n = \left[\begin{array}{cccc:c} & & & & 1 \\ & & & & 2 \\ & & I_{n-1} & & \vdots \\ & & & & n-1 \\ \hdashline 1 & 2 & \cdots & n-1 & n \end{array}\right]$$

If $A_n^{-1} = [b_{ij}]$ and $k = \dfrac{6}{n(n+1)(2n-5)}$,

$$b_{ii} = 1 - ki^2, \quad \text{if } 1 \le i \le n-1$$

$$b_{nn} = -k$$

$$b_{ij} = -kij, \qquad \text{if } i \ne j,\ 1 \le i \le n-1, \text{ and } 1 \le j \le n-1$$

$$b_{in} = b_{ni} = ki, \text{ if } 1 \le i \le n-1.$$

Determinant:

$$\det(A_n) = -\frac{1}{k}$$

Condition numbers (rounded):

n	$M(A_n)$	$P(A_n)$
2	8	6.854102
3	9	9.89898
4	14.4	6.531124
5	24	8.830950
6	35.2653	11.32624
7	48.4167	14
8	63.5152	16.83897
9	80.5846	19.83240
10	99.6364	22.97141

Reference: [1].

Example 3.17

$$A = \begin{bmatrix} a & 1 & & & & \\ 1 & a & 1 & & & \\ & 1 & a & 1 & & \\ & \cdots & \cdots & \cdots & \cdots & \\ & & & 1 & a & 1 \\ & & & & 1 & a \end{bmatrix}, \quad n \times n.$$

$A^{-1} = \frac{1}{b_n}[a_{ij}]$ is the n x n matrix defined by

$$a_{ij} = \begin{cases} b_{i-1}b_{n-i}, & \text{if } i = j \\ (-1)^{i+j} b_{i-1}b_{n-j}, & \text{if } j > i \\ a_{ji}, & \text{if } j < i \end{cases}$$

where

$$b_0 = 1$$

$$b_1 = a$$

$$b_k = ab_{k-1} - b_{k-2}, \quad k = 2, 3, \ldots, n.$$

Reference: [10].

Example 3.18

$$A = \begin{bmatrix} 2 & -1 & & & & \\ -1 & 2 & -1 & & & \\ & -1 & 2 & -1 & & \\ & \cdots & \cdots & \cdots & \cdots & \\ & & & -1 & 2 & -1 \\ & & & & -1 & 2 \end{bmatrix}, \quad n \times n.$$

$A^{-1} = \frac{1}{n+1} C$ where $C = [c_{ij}]$ is the n x n matrix defined by

$$c_{ij} = \begin{cases} i(n-i+1), & \text{if } i = j \\ c_{i,j-1} - i, & \text{if } j > i \\ c_{ji}, & \text{if } j < i \end{cases}$$

For example, when n = 4,

$$A = \begin{bmatrix} 2 & -1 & & \\ -1 & 2 & -1 & \\ & -1 & 2 & -1 \\ & & -1 & 2 \end{bmatrix} \qquad A^{-1} = \frac{1}{5}\begin{bmatrix} 4 & 3 & 2 & 1 \\ 3 & 6 & 4 & 2 \\ 2 & 4 & 6 & 3 \\ 1 & 2 & 3 & 4 \end{bmatrix}$$

Condition number:

$$P(A) \sim \frac{4n^2}{\pi^2} ,$$

Reference: [31], [43].

Example 3.19

Let n be an odd integer, and let A_{n+1} be the n+1 x n+1 matrix

$$A_{n+1} = \begin{bmatrix} 0 & x_1 & & & & \\ y_1 & 0 & x_2 & & & \\ & y_2 & 0 & x_3 & & \\ & \cdots & \cdots & \cdots & \cdots & \\ & & & y_{n-1} & 0 & x_n \\ & & & & y_n & 0 \end{bmatrix}$$

with $x_i \neq 0$ and $y_i \neq 0$, i = 1, 2, ..., n.

$$A_2^{-1} = \begin{bmatrix} 0 & \frac{1}{y_1} \\ \frac{1}{x_1} & 0 \end{bmatrix}$$

For $n \geq 3$,

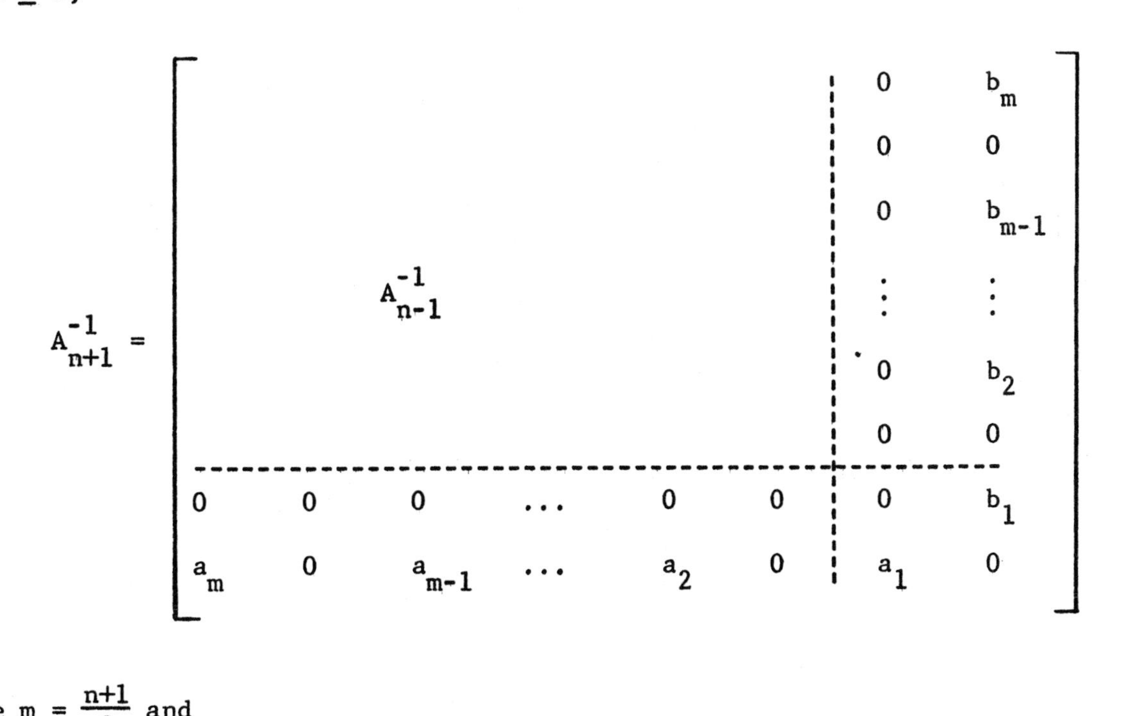

where $m = \frac{n+1}{2}$ and

$$a_k = \begin{cases} \dfrac{1}{x_n}, & \text{if } k = 1 \\[2ex] \dfrac{(-1)^{k+1}}{x_n} \displaystyle\prod_{j=1}^{k-1} \dfrac{y_{n-2j+1}}{x_{n-2j}}, & \text{if } k = 2, 3, \ldots, m. \end{cases}$$

and

$$b_k = \begin{cases} \dfrac{1}{y_n}, & \text{if } k = 1 \\[2ex] \dfrac{(-1)^{k+1}}{y_n} \displaystyle\prod_{j=1}^{k-1} \dfrac{x_{n-2j+1}}{y_{n-2j}}, & \text{if } k = 2, 3, \ldots, m. \end{cases}$$

Determinant:

$$\det(A_{n+1}) = (-1)^{(n+1)/2} \prod_{i=1}^{\frac{n+1}{2}} (x_{2i-1}y_{2i-1})$$

Reference: [12].

Example 3.20

$$A = \begin{bmatrix} (x+b) & 1 & & & & & \\ 1 & x & 1 & & & & \\ & 1 & x & 1 & & & \\ & & \cdots & \cdots & \cdots & & \\ & & & & 1 & x & 1 \\ & & & & & 1 & (x+a) \end{bmatrix}, \quad n \times n.$$

Inverse:

$A^{-1} = [c_{ij}]$ is the n x n matrix defined by

$$c_{ij} = c_{ji} = \frac{(-1)^{i+j} r_{j-1} s_{n-i}}{(x+a)r_{n-1} - r_{n-2}}, \quad \text{if } j \leq i,$$

where

$$r_0 = 1$$

$$r_1 = x + b$$

$$r_k = xr_{k-1} - r_{k-2}, \quad k = 2, 3, \ldots, n-1,$$

and

$$s_0 = 1$$

$$s_1 = x + a$$

$$s_k = xs_{k-1} - s_{k-2}, \quad k = 2, 3, \ldots, n-1.$$

For example, if

$$A = \begin{bmatrix} -3 & 1 & & & & \\ 1 & -2 & 1 & & & \\ & 1 & -2 & 1 & & \\ & \cdots & \cdots & \cdots & \cdots & \\ & & & 1 & -2 & 1 \\ & & & & 1 & -1 \end{bmatrix}$$

is an n x n matrix, then

$$A^{-1} = -\frac{1}{2} \begin{bmatrix} 1 & 1 & 1 & \cdots & 1 \\ 1 & 3 & 3 & \cdots & 3 \\ 1 & 3 & 5 & \cdots & 5 \\ \vdots & \vdots & \vdots & & \vdots \\ 1 & 3 & 5 & \cdots & 2n-1 \end{bmatrix}$$

Reference: [18, pp. 33-36], [43].

Example 3.21

$$\begin{bmatrix} (1-10^{-n}) & -1 & 1 & -1 \\ 1 & -1 & 1 & -1 \\ 1 & -1 & 0 & 0 \\ 1 & 0 & 0 & -1 \end{bmatrix} \begin{bmatrix} x_1 \\ x_2 \\ x_3 \\ x_4 \end{bmatrix} = \begin{bmatrix} -3 \\ -2 \\ -1 \\ -3 \end{bmatrix}$$

Solution:

$$x_1 = 10^n$$
$$x_2 = 10^n + 1$$
$$x_3 = 10^n + 2$$
$$x_4 = 10^n + 3$$

Reference: [52].

Example 3.22

$$\det\begin{bmatrix} -73 & 78 & 24 \\ 92 & 66 & 25 \\ -80 & 37 & 10 \end{bmatrix} = 1$$

$$\det\begin{bmatrix} -73 & 78 & 24 \\ 92 & 66 & 25 \\ -80 & 37 & 10.01 \end{bmatrix} = -118.94$$

$$\det\begin{bmatrix} -73 & 78 & 24 \\ 92.01 & 66 & 25 \\ -80 & 37 & 10 \end{bmatrix} = 2.08$$

$$\det\begin{bmatrix} -73 & 78.01 & 24 \\ 92 & 66 & 25 \\ -80 & 37 & 10 \end{bmatrix} = -28.20$$

Reference: [27], [56].

Example 3.23 (see also Example 3.6)

Let $A = [a_{ij}]$ be the $n \times n$ matrix defined by

$$a_{ij} = |i-j|.$$

Then

$$A^{-1} = -\frac{1}{2}\begin{bmatrix} (1-\frac{1}{n-1}) & -1 & & & & & -\frac{1}{n-1} \\ -1 & 2 & -1 & & & & \\ & -1 & 2 & -1 & & & \\ \cdots & \cdots & \cdots & \cdots & \cdots & \cdots & \cdots \\ & & -1 & 2 & -1 & & \\ & & & -1 & 2 & -1 & \\ -\frac{1}{n-1} & & & & -1 & (1-\frac{1}{n-1}) \end{bmatrix}.$$

$$\det A = (-1)^{n-1}2^{n-2}(n-1).$$

An interesting generalization of this matrix is also known [77, p. 32].

Reference: [77, p. 31].

Example 3.24

For arbitrary constants $a_1, a_2, \ldots, a_{n-1}$, let $A_n = A_n(a_1, a_2, \ldots, a_{n-1}) = [a_{ij}]$ denote the $n \times n$ matrix defined by

$$a_{ij} = \begin{cases} 1, & j \geq i, \\ a_j, & j < i. \end{cases}$$

For example,

$$A_4 = \begin{bmatrix} 1 & 1 & 1 & 1 \\ a_1 & 1 & 1 & 1 \\ a_1 & a_2 & 1 & 1 \\ a_1 & a_2 & a_3 & 1 \end{bmatrix}.$$

If we let $A_n^{-1} = [b_{ij}]$, then, for $n > 1$,

$$b_{ij} = \begin{cases} 1/(1-a_i), & i = j,\ i \neq n, \\ -1/(1-a_i), & j = i+1,\ i \neq n, \\ (a_{j-1}-a_j)/[(1-a_j)(1-a_{j-1})], & i = n,\ j \neq 1,\ n. \\ -a_1/(1-a_1), & i = n,\ j = 1, \\ 1/(1-a_{n-1}), & i = j = n, \\ 0, \ \text{otherwise}. \end{cases}$$

For $n = 1$, $A_1^{-1} = [1]$.

The determinant of A_n is given by

$$\det(A_n) = (1-a_1)(1-a_2) \ldots (1-a_{n-1}).$$

Hence A_n is singular if and only if $a_i = 1$ for some i.

Reference: [75].

<u>Example 3.25</u> (Combinatorial Matrix)

$$C = [y+\delta_{ij}x] = \begin{bmatrix} (x+y) & y & y & \cdots & y \\ y & (x+y) & y & \cdots & y \\ y & y & (x+y) & \cdots & y \\ \cdots & \cdots & \cdots & \cdots & \cdots \\ y & y & y & \cdots & (x+y) \end{bmatrix}$$

$$\det(C) = x^{n-1}(x+ny).$$

If $C^{-1} = [b_{ij}]$, then $\sum_{i,j} b_{ij} = \frac{n}{x+ny}$, and

$$b_{ij} = \frac{\delta_{ij}(x+ny)-y}{x(x+ny)}.$$

For example, if n = 3, x = 2, y = 1, then

$$C = \begin{bmatrix} 3 & 1 & 1 \\ 1 & 3 & 1 \\ 1 & 1 & 3 \end{bmatrix}, \quad C^{-1} = \frac{1}{10}\begin{bmatrix} 4 & -1 & -1 \\ -1 & 4 & -1 \\ -1 & -1 & 4 \end{bmatrix},$$

$$\det(C) = 20,$$

and

$$\sum_{i,j} b_{ij} = \frac{3}{5}.$$

Reference: [81, p. 36].

Example 3.26 (Cauchy's Matrix)

$$A = \left[\frac{1}{x_i+x_j}\right] = \begin{bmatrix} (x_1+y_1)^{-1} & (x_1+y_2)^{-1} & \cdots & (x_1+y_n)^{-1} \\ (x_2+y_1)^{-1} & (x_2+y_2)^{-1} & \cdots & (x_2+y_n)^{-1} \\ \cdots & \cdots & \cdots & \cdots \\ (x_n+y_1)^{-1} & (x_n+y_2)^{-1} & \cdots & (x_n+y_n)^{-1} \end{bmatrix}$$

$$\det(A) = \frac{\prod_{1\le i<j\le n} (x_j-x_i)(y_j-y_i)}{\prod_{1\le i,j\le n} (x_i+y_j)}$$

If $A^{-1} = [b_{ij}]$, then

$$b_{ij} = \frac{\prod_{1\le k\le n} (x_j+y_k)(x_k+y_i)}{(x_j+y_i)\left[\prod_{\substack{1\le k\le n \\ k\ne j}} (x_j-x_k)\right]\left[\prod_{\substack{1\le k\le n \\ k\ne i}} (y_i-y_k)\right]},$$

and $\sum_{i,j} b_{ij} = (x_1+x_2+\ldots+x_n) + (y_1+y_2+\ldots+y_n)$.

Note: The finite segments of the (infinite) Hilbert matrix are special examples of Cauchy's matrix. See example 3.8.

Reference: [81, p. 36].

Example 3.27 (Vandermonde's Matrix)

See Section 9 of Chapter II.

CHAPTER IV

TEST MATRICES: EIGENVALUES AND EIGENVECTORS OF REAL SYMMETRIC MATRICES

Example 4.1

$$\begin{bmatrix} 5 & 4 & 1 & 1 \\ 4 & 5 & 1 & 1 \\ 1 & 1 & 4 & 2 \\ 1 & 1 & 2 & 4 \end{bmatrix}$$

Eigenvalues:

$$\lambda_1 = 10$$
$$\lambda_2 = 5$$
$$\lambda_3 = 2$$
$$\lambda_4 = 1$$

Eigenvectors:

$$x_1 = \begin{bmatrix} 2 \\ 2 \\ 1 \\ 1 \end{bmatrix}, \quad x_2 = \begin{bmatrix} -1 \\ -1 \\ 2 \\ 2 \end{bmatrix}, \quad x_3 = \begin{bmatrix} 0 \\ 0 \\ -1 \\ 1 \end{bmatrix}, \quad x_4 = \begin{bmatrix} -1 \\ 1 \\ 0 \\ 0 \end{bmatrix}.$$

Reference: [49, pp. 54-55].

Example 4.2

$$\begin{bmatrix} 6 & 4 & 4 & 1 \\ 4 & 6 & 1 & 4 \\ 4 & 1 & 6 & 4 \\ 1 & 4 & 4 & 6 \end{bmatrix}$$

Eigenvalues:

$$\lambda_1 = 15$$
$$\lambda_2 = 5$$
$$\lambda_3 = 5$$
$$\lambda_4 = -1$$

Eigenvectors:

$$x_1 = \begin{bmatrix} 1 \\ 1 \\ 1 \\ 1 \end{bmatrix}, \quad x_2 = \begin{bmatrix} -1 \\ -1 \\ 1 \\ 1 \end{bmatrix}, \quad x_3 = \begin{bmatrix} -1 \\ 1 \\ -1 \\ 1 \end{bmatrix}, \quad x_4 = \begin{bmatrix} 1 \\ -1 \\ -1 \\ 1 \end{bmatrix}.$$

Note: For $\lambda_2 = \lambda_3$, we have a two-dimensional subspace of eigenvectors corresponding to this multiple eigenvalue. x_2 and x_3 are two orthogonal vectors from this subspace.

Reference: [49, pp. 53-54], [60, pp. 145-146].

Example 4.3

$$\begin{bmatrix} 2 & 1 & 3 & 4 \\ 1 & -3 & 1 & 5 \\ 3 & 1 & 6 & -2 \\ 4 & 5 & -2 & -1 \end{bmatrix}$$

Eigenvalues:

$$\lambda_1 \doteq -8.0285\ 7835$$
$$\lambda_2 \doteq 7.9329\ 0471$$
$$\lambda_3 \doteq 5.6688\ 6437$$
$$\lambda_4 \doteq -1.5731\ 9073$$

Eigenvectors:

$$x_1 \doteq \begin{bmatrix} 1.0000\ 0000 \\ 2.5014\ 6029 \\ -0.7577\ 3064 \\ -2.5642\ 1169 \end{bmatrix} \qquad x_2 \doteq \begin{bmatrix} 1.0000\ 0000 \\ 0.3778\ 1815 \\ 1.3866\ 2122 \\ 0.3488\ 0573 \end{bmatrix}$$

$$x_3 \doteq \begin{bmatrix} 1.0000\ 0000 \\ 0.9570\ 0150 \\ -1.4204\ 6822 \\ 1.7433\ 1690 \end{bmatrix} \qquad x_4 \doteq \begin{bmatrix} 1.0000\ 0000 \\ -0.9070\ 9211 \\ -0.3775\ 9122 \\ -0.3833\ 3124 \end{bmatrix}$$

Reference: [3], [8], [49, pp. 66-67], [74].

<u>Example 4.4</u>

$$\begin{bmatrix} 5 & 7 & 6 & 5 \\ 7 & 10 & 8 & 7 \\ 6 & 8 & 10 & 9 \\ 5 & 7 & 9 & 10 \end{bmatrix}$$

Characteristic polynomial:

$$P(\lambda) = 1 - 100\lambda + 146\lambda^2 - 35\lambda^3 + \lambda^4$$

Eigenvalues:

$$\lambda_1 \doteq 30.28868$$
$$\lambda_2 \doteq 3.85806$$
$$\lambda_3 \doteq 0.84311$$
$$\lambda_4 \doteq 0.01015$$

Inverse: Example 3.5

Reference: [27], [42, pp. 247-248], [49, p. 53].

Example 4.5

$$\begin{bmatrix} 0.81321 & -0.00013 & 0.00014 & 0.00011 & 0.00021 \\ -0.00013 & 0.93125 & 0.23567 & 0.41235 & 0.41632 \\ 0.00014 & 0.23567 & 0.18765 & 0.50632 & 0.30697 \\ 0.00011 & 0.41235 & 0.50632 & 0.27605 & 0.46322 \\ 0.00021 & 0.41632 & 0.30697 & 0.46322 & 0.41931 \end{bmatrix}$$

Tridiagonal form from Householder's method:

α_i	β_i
0.81321	...
0.57378	0.00030
1.33978	-0.48980
0.06519	-0.44013
-0.16450	0.17294

Eigenvalues:

$$\lambda_1 \doteq 1.67828$$

$$\lambda_2 \doteq 0.81321$$

$$\lambda_3 \doteq 0.41985$$

$$\lambda_4 \doteq 0.01521$$

$$\lambda_5 \doteq -0.29908$$

Reference: [67].

Example 4.6

$$\begin{bmatrix} 5 & 4 & 3 & 2 & 1 \\ 4 & 6 & 0 & 4 & 3 \\ 3 & 0 & 7 & 6 & 5 \\ 2 & 4 & 6 & 8 & 7 \\ 1 & 3 & 5 & 7 & 9 \end{bmatrix}$$

Eigenvalues:

$\lambda_1 \doteq 22.4068\ 7532$

$\lambda_2 \doteq 7.5137\ 24155$

$\lambda_3 \doteq 4.8489\ 50120$

$\lambda_4 \doteq 1.3270\ 45605$

$\lambda_5 \doteq -1.0965\ 95181$

Reference: [58], [68], [77].

Example 4.7

$$\begin{bmatrix} 10 & 1 & 2 & 3 & 4 \\ 1 & 9 & -1 & 2 & -3 \\ 2 & -1 & 7 & 3 & -5 \\ 3 & 2 & 3 & 12 & 1 \\ 4 & -3 & -5 & -1 & 15 \end{bmatrix}$$

Tridiagonal form from Householder's method:

α_i	β_i
9.295202 17754	...
11.626711 5560	0.749484 677741
10.960439 2078	-4.496268 20120
6.117647 05885	-2.157040 99085
15.000000 0000	7.141428 42854

Eigenvalues:

$\lambda_1 \doteq 1.655266\ 20775$

$\lambda_2 \doteq 6.994837\ 83064$

$\lambda_3 \doteq 9.363554\ 92016$

$\lambda_4 \doteq 15.808920\ 7645$

$\lambda_5 \doteq 19.175420\ 2773$

Reference: [37].

Example 4.8

$$\begin{bmatrix} 5 & 1 & -2 & 0 & -2 & 5 \\ 1 & 6 & -3 & 2 & 0 & 6 \\ -2 & -3 & 8 & -5 & -6 & 0 \\ 0 & 2 & -5 & 5 & 1 & -2 \\ -2 & 0 & -6 & 1 & 6 & -3 \\ 5 & 6 & 0 & -2 & -3 & 8 \end{bmatrix}$$

Eigenvalues:

$$\lambda_1 \doteq -1.598734\ 29358$$

$$\lambda_2 \doteq -1.598734\ 29346$$

$$\lambda_3 \doteq 4.455989\ 63847$$

$$\lambda_4 \doteq 4.455989\ 63855$$

$$\lambda_5 \doteq 16.142744\ 6551$$

$$\lambda_6 \doteq 16.142744\ 6553$$

Reference: [37].

Example 4.9

$$\begin{bmatrix} 1 & 2 & 3 & 0 & 1 & 2 \\ 2 & 4 & 5 & -1 & 0 & 3 \\ 3 & 5 & 6 & -2 & -3 & 0 \\ 0 & -1 & -2 & 1 & 2 & 3 \\ 1 & 0 & -3 & 2 & 4 & 5 \\ 2 & 3 & 0 & 3 & 5 & 6 \end{bmatrix}$$

Eigenvalues:

$$\lambda_1 \doteq 12.4113\ 3643$$

$$\lambda_2 \doteq 12.4113\ 3642$$

$$\lambda_3 \doteq 0.2849\ 864395$$

$$\lambda_4 \doteq 0.2849\ 864365$$

$$\lambda_5 \doteq -1.6963\ 22849$$

$$\lambda_6 \doteq -1.6963\ 22851$$

Reference: [58], [68], [77].

Example 4.10 (Rosser, *et al*.)

$$A = \begin{bmatrix} 611 & 196 & -192 & 407 & -8 & -52 & -49 & 29 \\ 196 & 899 & 113 & -192 & -71 & -43 & -8 & -44 \\ -192 & 113 & 899 & 196 & 61 & 49 & 8 & 52 \\ 407 & -192 & 196 & 611 & 8 & 44 & 59 & -23 \\ -8 & -71 & 61 & 8 & 411 & -599 & 208 & 208 \\ -52 & -43 & 49 & 44 & -599 & 411 & 208 & 208 \\ -49 & -8 & 8 & 59 & 208 & 208 & 99 & -911 \\ 29 & -44 & 52 & -23 & 208 & 208 & -911 & 99 \end{bmatrix}$$

Tridiagonal form from Lanczos' Method for $10^{-3}A$:

α_i	β_i
0.899	....
0.1086629633	0.096939
0.7859177671	0.039517948848
-0.7935214279	0.4088977136
0.003963315517	0.0520498144977
1.0160663075	0.004021099703
1.0199110708	$0.1070421101 \times 10^{-8}$
1.0000000030	$0.7048359779 \times 10^{-10}$

Eigenvalues:

$$\lambda_1 = 10\sqrt{10405} \doteq 1020.04901843$$

$$\lambda_2 = 1020$$

$$\lambda_3 = 510 + 100\sqrt{26} \doteq 1019.90195136$$

$$\lambda_4 = 1000$$

$$\lambda_5 = 1000$$

$$\lambda_6 = 510 - 100\sqrt{26} \doteq 0.09804864072$$

$$\lambda_7 = 0$$

$$\lambda_8 = -10\sqrt{10405} \doteq -1020.04901843$$

Eigenvectors:

$$x_1 = \begin{bmatrix} 2 \\ 1 \\ 1 \\ 2 \\ 102 - \sqrt{10405} \\ 102 - \sqrt{10405} \\ -204 + 2\sqrt{10405} \\ -204 + 2\sqrt{10405} \end{bmatrix}, \quad x_2 = \begin{bmatrix} 1 \\ -2 \\ -2 \\ 1 \\ 2 \\ -2 \\ 1 \\ -1 \end{bmatrix}, \quad x_3 = \begin{bmatrix} 2 \\ -1 \\ 1 \\ -2 \\ 5 + \sqrt{26} \\ -5 - \sqrt{26} \\ -10 - 2\sqrt{26} \\ 10 + 2\sqrt{26} \end{bmatrix}, \quad x_4 = \begin{bmatrix} 1 \\ -2 \\ -2 \\ 1 \\ -2 \\ 2 \\ -1 \\ 1 \end{bmatrix}$$

$$x_5 = \begin{bmatrix} 7 \\ 14 \\ -14 \\ -7 \\ -2 \\ -2 \\ -1 \\ -1 \end{bmatrix}, \quad x_6 = \begin{bmatrix} 2 \\ -1 \\ 1 \\ -2 \\ 5 - \sqrt{26} \\ -5 + \sqrt{26} \\ -10 + 2\sqrt{26} \\ 10 - 2\sqrt{26} \end{bmatrix}, \quad x_7 = \begin{bmatrix} 1 \\ 2 \\ -2 \\ -1 \\ 14 \\ 14 \\ 7 \\ 7 \end{bmatrix}, \quad x_8 = \begin{bmatrix} 2 \\ 1 \\ 1 \\ 2 \\ 102 + \sqrt{10405} \\ 102 + \sqrt{10405} \\ -204 - 2\sqrt{10405} \\ -204 - 2\sqrt{10405} \end{bmatrix}$$

Note: For $\lambda_4 = \lambda_5$ we have a two-dimensional subspace of eigenvectors corresponding to this multiple eigenvalue. x_4 and x_5 are two orthogonal vectors from this subspace.

Reference: [47].

Example 4.11

$$
\begin{bmatrix}
5 & 2 & 1 & 1 & & & & & & & \\
2 & 6 & 3 & 1 & 1 & & & & & & \\
1 & 3 & 6 & 3 & 1 & 1 & & & & & \\
1 & 1 & 3 & 6 & 3 & 1 & 1 & & & & \\
 & 1 & 1 & 3 & 6 & 3 & 1 & 1 & & & \\
 & & 1 & 1 & 3 & 6 & 3 & 1 & 1 & & \\
 & & & 1 & 1 & 3 & 6 & 3 & 1 & 1 & \\
 & & & & 1 & 1 & 3 & 6 & 3 & 1 & 1 \\
 & & & & & 1 & 1 & 3 & 6 & 3 & 1 \\
 & & & & & & 1 & 1 & 3 & 6 & 2 \\
 & & & & & & & 1 & 1 & 2 & 5
\end{bmatrix}
$$

Eigenvalues:

$$\lambda_1 \doteq 14.94181\ 93276\ 76382$$

$$\lambda_2 \doteq 12.19615\ 24227\ 06632$$

$$\lambda_3 \doteq 8.82842\ 71247\ 461900$$

$$\lambda_4 \doteq 6.00000\ 00000\ 000000$$

$$\lambda_5 \doteq 4.40664\ 99006\ 731521$$

$$\lambda_6 \doteq 4.12924\ 84841\ 890931$$

$$\lambda_7 \doteq 4.00000\ 00000\ 000000$$

$$\lambda_8 \doteq 4.00000\ 00000\ 000000$$

$$\lambda_9 \doteq 3.17157\ 28752\ 538100$$

$$\lambda_{10} \doteq 1.80384\ 75772\ 933680$$

$$\lambda_{11} \doteq 0.52228\ 22874\ 6137256$$

Reference: [49, pp. 78-79], [66].

Example 4.12

0.25000	0.06675	0.04000	0.02475	0.07050	0.06375	0.06925	0.02050	0.03600	-0.01025	-0.00175	0.02750	0.02300	0.00200
0.06675	0.25000	0.10400	0.07475	0.03625	0.11675	0.11050	0.06225	0.05100	0.03250	0.02400	0.03600	0.06350	0.05300
0.04000	0.10400	0.25000	0.14575	0.03725	0.07175	0.07800	0.12200	0.11275	0.09375	0.10175	0.09600	0.14300	0.11550
0.02475	0.07475	0.14575	0.25000	0.05375	0.07000	0.05225	0.12800	0.12475	0.10550	0.13000	0.14575	0.13975	0.13375
0.07050	0.03625	0.03725	0.05375	0.25000	0.04575	0.05750	0.05700	0.05050	0.01475	0.04500	0.07150	0.05300	0.01600
0.06375	0.11675	0.07175	0.07000	0.04575	0.25000	0.08625	0.08800	0.07150	0.04850	0.03200	0.04475	0.03300	0.04500
0.06925	0.11050	0.07800	0.05225	0.05750	0.08625	0.25000	0.08725	0.06950	0.03725	0.04025	0.04300	0.04075	0.01450
0.02050	0.06225	0.12200	0.12800	0.05700	0.08800	0.08725	0.25000	0.14100	0.13275	0.15550	0.13050	0.11825	0.09125
0.03600	0.05100	0.11275	0.12475	0.05050	0.07150	0.06950	0.14100	0.25000	0.07425	0.10750	0.09175	0.10725	0.08225
-0.01025	0.03250	0.09375	0.10550	0.01475	0.04850	0.03725	0.13275	0.07425	0.25000	0.15500	0.09625	0.09950	0.09425
-0.00175	0.02400	0.10175	0.13000	0.04500	0.03200	0.04025	0.15550	0.10750	0.15500	0.25000	0.13350	0.14850	0.13050
0.02750	0.03600	0.09600	0.14575	0.07150	0.04475	0.04300	0.13050	0.09175	0.09625	0.13350	0.25000	0.11100	0.10075
0.02300	0.06350	0.14300	0.13975	0.05300	0.03300	0.04075	0.11825	0.10725	0.09950	0.14850	0.11100	0.25000	0.14325
0.00200	0.05300	0.11550	0.13375	0.01600	0.04500	0.01450	0.09125	0.08225	0.09425	0.13050	0.10075	0.14325	0.25000

Tridiagonal form from Givens' Method:

α_i	β_i
0.25000 0000	.....
0.76849 1173	0.15366 0746
0.91955 6756	0.46726 0328
0.23093 8895	0.11925 6498
0.13305 3788	0.08076 3539
0.22254 9575	0.03394 7196
0.11612 7856	0.03609 0904
0.12033 9373	0.03502 2375
0.12371 9912	0.02915 7561
0.12856 1407	0.03745 3705
0.10776 8089	0.01609 0599
0.13703 9203	0.02382 6467
0.13805 7030	0.02946 8449
0.10379 6943	0.00764 6394

Eigenvalues:

$\lambda_1 \doteq 1.33403\ 48369\ 565070$

$\lambda_2 \doteq 0.46276\ 62026\ 942646$

$\lambda_3 \doteq 0.26773\ 32979\ 540935$

$\lambda_4 \doteq 0.23163\ 94839\ 784262$

$\lambda_5 \doteq 0.17735\ 63338\ 214251$

$\lambda_6 \doteq 0.17130\ 75618\ 005174$

$\lambda_7 \doteq 0.16632\ 46020\ 889680$

$\lambda_8 \doteq 0.14342\ 28761\ 463870$

$\lambda_9 \doteq 0.12278\ 75231\ 815480$

$\lambda_{10} \doteq 0.10321\ 57624\ 309478$

$\lambda_{11} \doteq 0.09720\ 92161\ 618017$

$\lambda_{12} \doteq 0.08422\ 52705\ 646369$

$\lambda_{13} \doteq 0.07359\ 71188\ 537465$

$\lambda_{14} \doteq 0.06437\ 99133\ 667302$

Eigenvectors:

We give the first four eigenvectors because the later components of these vectors are very small. The other eigenvectors are quite normal in form.

x_1	x_2	x_3	x_4
0.12180621	0.72220468	0.49474263	-0.34701603
0.85930971	1.00000000	0.05709601	0.04146403
1.00000000	-0.89179298	-0.22388762	0.06647838
0.10864611	-0.50225957	1.00000000	-0.54593352
0.00731255	-0.12487537	0.78617724	-0.10289858
0.00022356	-0.01793071	0.73992634	1.00000000
0.00000663	-0.00188653	0.18686459	0.34864816
0.00000019	-0.00019439	0.04640116	0.11940974
0.00000001	-0.00001692	0.01011081	0.03703434
0.00000000	-0.00000190	0.00275399	0.01375120
0.00000000	-0.00000009	0.00028519	0.00188770
0.00000000	-0.00000000	0.00005483	0.00052743
0.00000000	0.00000000	0.00001250	0.00016690
0.00000000	0.00000000	0.00000059	0.00000998

Reference: [9], [63], [66], [67].

Example 4.13 (Hilbert Matrix)

Let $A_n = \left[a_{ij}^{(n)}\right]$ be the n x n matrix defined by

$$a_{ij}^{(n)} = \frac{1}{i+j-1}, \quad i,j = 1, 2, \ldots, n.$$

$$A_n = \begin{bmatrix} 1 & \frac{1}{2} & \frac{1}{3} & \cdots & \frac{1}{n} \\ \frac{1}{2} & \frac{1}{3} & \frac{1}{4} & \cdots & \frac{1}{n+1} \\ \frac{1}{3} & \frac{1}{4} & \frac{1}{5} & \cdots & \frac{1}{n+2} \\ \cdots & \cdots & \cdots & \cdots & \cdots \\ \frac{1}{n} & \frac{1}{n+1} & \frac{1}{n+2} & \cdots & \frac{1}{2n-1} \end{bmatrix}$$

Eigenvalues and Eigenvectors:

The eigenvalues and eigenvectors of A_n, n = 3, 4, ..., 10, are given on the following pages. In addition, we give the eigenvalue of largest magnitude and the corresponding eigenvector for A_2 and A_{20}.

A_2

Eigenvalue	Eigenvector
1.26759 188	1.00000 000
	0.53518 376

A_{20}

Eigenvalue	Eigenvector
1.90713 472	1.00000 000
	0.63153 893
	0.48170 552
	0.39577 939
	0.33864 052
	0.29732 839
	0.26579 806
	0.24080 108
	0.22041 627
	0.20342 569
	0.18901 536
	0.17661 823
	0.16582 577
	0.15633 540
	0.14791 772
	0.14039 536
	0.13362 876
	0.12750 652
	0.12193 851
	0.11685 095

Inverse: Example 3.8

Reference: [19], [20], [77, p. 30].

ORDER OF MATRIX = 3

EIGENVALUES	EIGENVECTORS	EIGENVALUES	EIGENVECTORS
1.40831 89271 23654(-00)	1.00000 00000 00000(-00)	1.22327 06585 39058(-01)	-8.43517 43276 29785(-01)
	5.56032 55563 05693(-01)		8.13998 17376 62614(-01)
	3.90907 94792 51080(-01)		1.00000 00000 00000(-00)
2.68734 03557 73529(-03)	-1.78857 98493 23438(-01)		
	1.00000 00000 00000(-00)		
	-9.64868 00204 55515(-01)		

ORDER OF MATRIX = 4

EIGENVALUES	EIGENVECTORS	EIGENVALUES	EIGENVECTORS
1.50021 42800 59243(-00)	1.00000 00000 00000(-00)	1.69141 22022 14500(-01)	1.00000 00000 00000(-00)
	5.70172 08366 32358(-01)		-6.36518 90190 07507(-01)
	4.06778 98802 75292(-01)		-8.75450 79607 67703(-01)
	3.18140 96887 37940(-01)		-8.83129 58721 03381(-01)
6.73827 36057 60748(-03)	-2.41517 71638 15848(-01)	9.67023 04022 58689(-05)	3.68876 82614 14105(-02)
	1.00000 00000 00000(-00)		-4.15349 28778 03112(-01)
	-1.35093 31925 07654(-01)		1.00000 00000 00000(-00)
	-8.60314 35862 04442(-01)		-6.50171 21973 36798(-01)

ORDER OF MATRIX = 5

EIGENVALUES	EIGENVECTORS	EIGENVALUES	EIGENVECTORS
1.56705 06910 98231(-00)	1.00000 00000 00000(-00)	2.08534 21861 10133(-01)	1.00000 00000 00000(-00)
	5.80566 92249 80478(-01)		-4.58425 80576 61740(-01)
	4.18800 95256 90560(-01)		-7.05925 82907 15063(-01)
	3.30061 05409 17674(-01)		-7.37537 92074 31147(-01)
	2.73258 24401 62320(-01)		-7.12798 94314 80946(-01)

ORDER OF MATRIX = 5 (CONT.)

EIGENVALUES	EIGENVECTORS	EIGENVALUES	EIGENVECTORS
1.14074 91623 41981(-02)	-2.95833 43954 91379(-01)	3.05898 04015 11917(-04)	7.06702 26210 87525(-02)
	1.00000 00000 00000(-00)		-6.48336 02593 66261(-01)
	1.66348 46563 67509(-01)		1.00000 00000 00000(-00)
	-4.27528 04665 91248(-01)		3.49178 63233 06241(-01)
	-7.80543 77407 62442(-01)		-8.35542 93387 42830(-01)
3.28792 87721 71863(-06)	-8.04735 96573 69526(-03)		
	1.52103 86654 52718(-01)		
	-6.59762 08136 21921(-01)		
	1.00000 00000 00000(-00)		
	-4.90419 53143 50719(-01)		

ORDER OF MATRIX = 6

EIGENVALUES	EIGENVECTORS	EIGENVALUES	EIGENVECTORS
1.61889 98589 24339(-00)	1.00000 00000 00000(-00)	2.42360 87057 52096(-01)	1.00000 00000 00000(-00)
	5.88628 54342 55432(-01)		-3.43477 76103 67806(-01)
	4.28327 28442 89561(-01)		-5.95389 61269 85598(-01)
	3.39661 89183 87095(-01)		-6.42274 99431 02546(-01)
	2.82523 58794 21492(-01)		-6.31671 46805 61395(-01)
	2.42337 81112 28495(-01)		-6.03204 01490 85321(-01)
1.63215 21319 87582(-02)	-3.44477 74040 00321(-01)	6.15748 35418 26577(-04)	-1.15089 16226 58221(-01)
	1.00000 00000 00000(-00)		9.07815 69634 66591(-01)
	3.31669 05639 78445(-01)		-9.90373 92462 04362(-01)
	-1.90443 48397 72404(-01)		-7.71318 49997 79162(-01)
	-5.19908 55937 27446(-01)		8.69902 39991 00457(-02)
	-7.20650 57788 73129(-01)		1.00000 00000 00000(-00)
1.25707 57122 62519(-05)	1.84443 82298 42188(-02)	1.08279 94845 65550(-07)	1.80948 25414 40515(-03)
	-2.97466 27961 49800(-01)		-5.16182 53594 24858(-02)
	1.00000 00000 00000(-00)		3.48907 75253 55039(-01)
	-7.34137 29699 37382(-01)		-9.06717 68457 84127(-01)
	-7.30764 18529 36246(-01)		1.00000 00000 00000(-00)
	7.59856 90405 64665(-01)		-3.93741 11149 37020(-01)

ORDER OF MATRIX = 7

EIGENVALUES	EIGENVECTORS
1.66088 53389 26931(-00)	1.00000 00000 00000(-00)
	5.95122 63106 51334(-01)
	4.36126 49735 70395(-01)
	3.47622 28057 85343(-01)
	2.90284 56422 45982(-01)
	2.49777 30606 63215(-01)
	2.19495 43192 32110(-01)
2.12897 54908 32795(-02)	-3.88993 12692 28422(-01)
	1.00000 00000 00000(-00)
	4.40423 18707 70565(-01)
	-3.43693 09768 67312(-02)
	-3.48496 99421 85792(-01)
	-5.48611 13060 83639(-01)
	-6.74584 79000 81032(-01)
2.93863 68145 92969(-05)	2.54375 48028 50871(-02)
	-3.62466 87695 55796(-01)
	1.00000 00000 00000(-00)
	-3.18671 25647 34205(-01)
	-7.90476 81764 75158(-01)
	-2.94027 65031 95887(-01)
	7.64617 63467 28869(-01)
3.49389 86059 91218(-09)	3.59098 91821 95847(-04)
	-1.44149 72273 50558(-02)
	1.39474 73803 77168(-01)
	-5.44035 08875 84887(-01)
	1.00000 00000 00000(-00)
	-8.65947 69018 12042(-01)
	2.84831 36565 59360(-01)
2.71920 19814 93452(-01)	1.00000 00000 00000(-00)
	-2.61651 87231 55985(-01)
	-5.15876 57109 07207(-01)
	-5.73403 92060 79247(-01)
	-5.72924 00712 20064(-01)
	-5.52870 16536 01293(-01)
	-5.26499 39668 34787(-01)
1.00858 76107 70142(-03)	-1.42730 60664 90882(-01)
	1.00000 00000 00000(-00)
	-8.08037 82159 87594(-01)
	-8.76415 71705 77919(-01)
	-3.24986 39912 94217(-01)
	3.46758 27944 15234(-01)
	9.67685 27364 04640(-01)
4.85676 33615 74250(-07)	-3.82934 35999 28926(-03)
	9.58555 22029 61430(-02)
	-5.40843 92603 82367(-01)
	1.00000 00000 00000(-00)
	-2.70501 09557 80551(-01)
	-8.43244 13447 80657(-01)
	5.65766 58320 00075(-01)

ORDER OF MATRIX = 8

EIGENVALUES	EIGENVECTORS	EIGENVALUES	EIGENVECTORS
1.69593 89969 21949(-00)	1.00000 00000 00000(-00)	2.98125 21131 69307(-01)	1.00000 00000 00000(-00)
	6.00504 24575 79538(-01)		-1.99641 07668 62729(-01)
	4.42671 55401 19186(-01)		-4.55006 08109 55082(-01)
	3.54370 44699 96978(-01)		-5.20378 81437 80070(-01)
	2.96918 57844 45071(-01)		-5.27565 95376 14758(-01)
	2.56180 92948 69805(-01)		-5.13976 52248 24774(-01)
	2.25629 36880 82276(-01)		-4.92889 11462 14828(-01)
	2.01790 18703 79183(-01)		-4.69617 42309 87346(-01)
2.62128 43578 11905(-02)	-4.30353 53605 31482(-01)	1.46768 81177 41867(-03)	-1.57264 17782 81082(-01)
	1.00000 00000 00000(-00)		1.00000 00000 00000(-00)
	5.19670 85157 87076(-01)		-6.11217 36807 18912(-01)
	7.95547 21960 88871(-02)		-8.34786 09555 82616(-01)
	-2.23317 97982 52556(-01)		-5.01055 24897 20307(-01)
	-4.23090 40968 60939(-01)		-2.04888 90803 70457(-02)
	-5.53618 29278 79063(-01)		4.52548 32316 57058(-01)
	-6.38180 47813 30543(-01)		8.67560 91295 41801(-01)
5.43694 33697 49942(-05)	3.30983 85076 81348(-02)	1.29433 20918 72811(-06)	-6.57331 00511 27815(-03)
	-4.26438 76153 95457(-01)		1.47989 76372 18060(-01)
	1.00000 00000 00000(-00)		-7.22416 37481 59994(-01)
	-5.12022 50480 36401(-02)		1.00000 00000 00000(-00)
	-6.72019 76374 56363(-01)		2.12540 06545 33205(-01)
	-5.85721 70581 51348(-01)		-7.21939 02421 98936(-01)
	-2.88541 78933 63593(-02)		-6.07829 90639 21476(-01)
	7.65890 29707 01414(-01)		7.04255 03469 70941(-01)
1.79887 37458 17577(-08)	-9.20944 89718 87311(-04)	1.11153 89663 72442(-10)	-6.86103 92145 12811(-05)
	3.30647 67571 34515(-02)		3.68787 70518 27661(-03)
	-2.77763 32839 45767(-01)		-4.82672 54524 49843(-02)
	8.69385 31714 42472(-01)		2.61713 39967 61041(-01)
	-9.95882 21590 46623(-01)		-7.05747 34717 96188(-01)
	-1.31307 33940 75290(-01)		1.00000 00000 00000(-00)
	1.00000 00000 00000(-00)		-7.12509 13818 01248(-01)
	-4.97329 08134 30048(-01)		2.01241 83438 37764(-01)

ORDER OF MATRIX = 9

EIGENVALUES	EIGENVECTORS	EIGENVALUES	EIGENVECTORS
1.72588 26609 01847(-00)	1.00000 00000 00000(-00)	3.21633 12229 92068(-01)	1.00000 00000 00000(-00)
	6.05062 73643 51117(-01)		-1.50563 00038 85823(-01)
	4.48271 58106 99431(-01)		-4.06370 51543 62231(-01)
	3.60192 03013 69706(-01)		-4.77760 59330 52471(-01)
	3.02681 26027 11120(-01)		-4.90983 21999 98590(-01)
	2.61776 14674 05719(-01)		-4.82553 42744 47961(-01)
	2.31016 02171 39396(-01)		-4.65723 08579 25155(-01)
	2.06956 70822 18468(-01)		-4.45945 07919 47498(-01)
	1.87577 43844 56729(-01)		-4.25620 54046 16327(-01)
3.10389 25781 26633(-02)	-4.69215 60414 24556(-01)	1.97893 38602 15924(-03)	-1.70653 85569 90212(-01)
	1.00000 00000 00000(-00)		1.00000 00000 00000(-00)
	5.81298 48942 56370(-01)		-4.64717 14467 22524(-01)
	1.68338 31445 98111(-01)		-7.75951 09562 54300(-01)
	-1.25629 34840 26744(-01)		-5.78327 97815 76324(-01)
	-3.25116 29771 00839(-01)		-2.21347 44112 47116(-01)
	-4.59281 69753 28175(-01)		1.53939 04201 98640(-01)
	-5.49127 85754 13251(-01)		4.96040 77413 22528(-01)
	-6.08710 62023 19202(-01)		7.89930 71224 23032(-01)
8.75808 50514 59757(-05)	4.13830 29491 86680(-02)	2.67301 34105 99414(-06)	-1.05472 67848 64242(-02)
	-4.90024 66072 79400(-01)		2.17325 36475 07308(-01)
	1.00000 00000 00000(-00)		-9.40220 91065 07029(-01)
	1.41406 98631 98621(-01)		1.00000 00000 00000(-00)
	-5.29978 89637 04739(-01)		5.70068 66549 30026(-01)
	-6.54975 57014 80830(-01)		-4.36808 34364 41685(-01)
	-3.71913 95311 21723(-01)		-8.57092 52955 24884(-01)
	1.44275 27836 59178(-01)		-3.92465 30179 95107(-01)
	7.66821 37097 76299(-01)		8.60212 23900 26319(-01)
5.38561 33485 22494(-08)	-1.57026 36778 04265(-03)	6.46090 54226 38582(-10)	1.86895 58009 44782(-04)
	5.13643 34130 24127(-02)		-9.11142 60410 63340(-03)
	-3.83966 02990 96905(-01)		1.06097 02731 82322(-01)
	1.00000 00000 00000(-00)		-4.88842 63971 64796(-01)
	-7.07732 14392 12621(-01)		1.00000 00000 00000(-00)
	-6.33233 77263 71795(-01)		-7.08369 80999 49995(-01)
	4.61853 32026 77920(-01)		-4.70608 86461 34285(-01)
	8.23643 49129 50087(-01)		9.44390 21369 08030(-01)
	-6.11750 86298 36100(-01)		-3.73891 52081 99850(-01)

ORDER OF MATRIX = 9 (CONT.)

EIGENVALUES	EIGENVECTORS
3.49967 64029 11493(-12)	1.36620 49070 13275(-05)
	-9.47535 57566 95441(-04)
	1.61058 61751 07205(-02)
	-1.15383 19398 86663(-01)
	4.24473 99502 14224(-01)
	-8.68875 53580 11983(-01)
	1.00000 00000 00000(-00)
	-6.05131 38110 13442(-01)
	1.49754 18355 81289(-01)

ORDER OF MATRIX = 10

EIGENVALUES	EIGENVECTORS	EIGENVALUES	EIGENVECTORS
1.75191 96702 65178(-00)	1.00000 00000 00000(-00)	3.42929 54848 35091(-01)	1.00000 00000 00000(-00)
	6.08991 91436 96503(-01)		-1.10465 17177 43785(-01)
	4.53138 29895 94215(-01)		-3.66282 37964 87492(-01)
	3.65286 01340 21510(-01)		-4.42425 91767 28277(-01)
	3.07753 04744 55016(-01)		-4.60536 94518 17078(-01)
	2.66725 18429 30508(-01)		-4.56341 25730 07357(-01)
	2.35801 30798 24843(-01)		-4.43036 22451 70023(-01)
	2.11563 96395 15401(-01)		-4.26171 31611 63669(-01)
	1.92005 12818 61191(-01)		-4.08260 80801 73888(-01)
	1.75860 03439 31029(-01)		-3.90483 78675 18817(-01)
3.57418 16271 63924(-02)	-5.06044 64866 39978(-01)	2.53089 07686 70038(-03)	-1.83100 74913 05559(-01)
	1.00000 00000 00000(-00)		1.00000 00000 00000(-00)
	6.31415 38757 61648(-01)		-3.50413 96416 78791(-01)
	2.40699 24192 57739(-01)		-7.15015 97273 02427(-01)
	-4.58618 75111 07758(-02)		-6.09646 75122 58232(-01)
	-2.45040 59873 37491(-01)		-3.39133 24904 62540(-01)
	-3.82178 19752 80346(-01)		-3.36629 40180 83149(-02)
	-4.76401 86973 73599(-01)		2.55456 55324 15149(-01)
	-5.40836 48066 89185(-01)		5.10538 49054 81802(-01)
	-5.84363 76347 92411(-01)		7.27979 70808 98943(-01)

ORDER OF MATRIX = 10 (CONT.)

EIGENVALUES	EIGENVECTORS	EIGENVALUES	EIGENVECTORS
1.28749 61427 63771(-04)	5.02691 93708 19130(-02)	4.72968 92931 82348(-06)	1.34197 07196 31349(-02)
	-5.53715 19652 43531(-01)		-2.56349 00053 68558(-01)
	1.00000 00000 00000(-00)		1.00000 00000 00000(-00)
	2.90771 16075 67017(-01)		-8.26254 49309 85651(-01)
	-3.93565 04576 91000(-01)		-7.30398 87720 04190(-01)
	-6.42818 76489 72120(-01)		9.79799 70771 83957(-02)
	-5.30534 27508 17633(-01)		6.90035 19449 49648(-01)
	-1.92318 51809 90450(-01)		7.09559 50102 98084(-01)
	2.64461 64931 15005(-01)		1.52328 09663 07994(-01)
	7.68640 20927 86739(-01)		-8.64618 11259 32675(-01)
1.22896 77387 51175(-07)	-2.20211 62655 20837(-03)	2.14743 88173 50479(-09)	3.29136 10483 09510(-04)
	6.65308 34126 35937(-02)		-1.47645 74515 12742(-02)
	-4.50423 02451 95166(-01)		1.55750 54872 45585(-01)
	1.00000 00000 00000(-00)		-6.25397 96321 29557(-01)
	-3.97406 18263 63392(-01)		1.00000 00000 00000(-00)
	-7.52617 34673 95069(-01)		-2.34914 01869 53356(-01)
	-5.19153 37204 50187(-02)		-7.83112 53616 02477(-01)
	6.56558 24094 19764(-01)		1.07843 33964 72780(-01)
	5.76366 49789 61755(-01)		8.66944 29913 73357(-01)
	-6.46988 57273 97804(-01)		-4.72962 81021 99039(-01)
2.26674 67477 62926(-11)	-3.73254 77290 76785(-05)	1.09315 38193 79666(-13)	2.71471 31336 04098(-06)
	2.37327 10245 65077(-03)		-2.36061 26295 90383(-04)
	-3.64962 96430 35628(-02)		5.05289 73867 16890(-03)
	2.29661 97587 73360(-01)		-4.61160 40049 98925(-02)
	-6.96603 89892 21241(-01)		2.20661 51772 89104(-01)
	1.00000 00000 00000(-00)		-6.08176 78395 43368(-01)
	-3.68315 31552 82386(-01)		1.00000 00000 00000(-00)
	-6.99899 14802 67557(-01)		-9.68158 87951 22191(-01)
	8.48690 27725 93223(-01)		5.09073 58516 71383(-01)
	-2.79403 11288 36095(-01)		-1.12104 94021 47474(-01)

Example 4.14

$$\begin{bmatrix} n & n-1 & n-2 & \cdots & 2 & 1 \\ n-1 & n-1 & n-2 & \cdots & 2 & 1 \\ n-2 & n-2 & n-2 & \cdots & 2 & 1 \\ \cdots & \cdots & \cdots & \cdots & \cdots & \cdots \\ 2 & 2 & 2 & \cdots & 2 & 1 \\ 1 & 1 & 1 & \cdots & 1 & 1 \end{bmatrix}$$

Eigenvalues:

$$\lambda_i = \frac{1}{2}\left[1 - \cos\frac{(2i-1)\pi}{2n+1}\right]^{-1}, \quad i = 1, 2, \ldots, n.$$

Characteristic polynomial for n = 12:

$$P(\lambda) = \lambda^{12} - 78\lambda^{11} + 1001\lambda^{10} - 5005\lambda^{9} + 12870\lambda^{8} - 19448\lambda^{7} + 18564\lambda^{6} - 11628\lambda^{5} + 4845\lambda^{4} - 1330\lambda^{3} + 231\lambda^{2} - 23\lambda + 1$$

Inverse: Example 3.12

Reference: [24].

Example 4.15

$$A_n = \left[\begin{array}{cccc|c} & & & & 1 \\ & & & & 2 \\ & I_{n-1} & & & \vdots \\ & & & & n-1 \\ \hline 1 & 2 & \cdots & n-1 & n \end{array}\right]$$

Eigenvalues:

$$\lambda_1 = \lambda_2 = \cdots = \lambda_{n-2} = 1$$

λ_{n-1} and λ_n are the roots of $\lambda^2 - (n+1)\lambda + \det(A_n) = 0$ where

$$\det(A_n) = -\frac{n(n+1)(2n-5)}{6}$$

Inverse: Example 3.16

Reference: [1].

Example 4.16

$$\begin{bmatrix} 5 & -4 & 1 & & & & \\ -4 & 6 & -4 & 1 & & & \\ 1 & -4 & 6 & -4 & 1 & & \\ \cdots & \cdots & \cdots & \cdots & \cdots & \cdots & \cdots \\ & & 1 & -4 & 6 & -4 & 1 \\ & & & 1 & -4 & 6 & -4 \\ & & & & 1 & -4 & 5 \end{bmatrix}, \quad n \times n.$$

Eigenvalues:

$$\lambda_k = 16 \sin^4 \left(\frac{k\pi}{2(n+1)} \right), \qquad k = 1, 2, \ldots, n.$$

Reference: [18, p. 20], [58].

Example 4.17

$$A = \begin{bmatrix} (B+I_n) & 16I_n & -I_n & & & & \\ 16I_n & B & 16I_n & -I_n & & & \\ -I_n & 16I_n & B & 16I_n & -I_n & & \\ \cdots & \cdots & \cdots & \cdots & \cdots & \cdots & \cdots \\ & & & -I_n & 16I_n & B & 16I_n \\ & & & & -I_n & 16I_n & (B+I_n) \end{bmatrix}, \quad n^2 \times n^2,$$

where

$$B = \begin{bmatrix} -59 & 16 & -1 & & & & \\ 16 & -60 & 16 & -1 & & & \\ -1 & 16 & -60 & 16 & -1 & & \\ \cdots & \cdots & \cdots & \cdots & \cdots & \cdots & \cdots \\ & & & -1 & 16 & -60 & 16 \\ & & & & -1 & 16 & -59 \end{bmatrix}, \quad n \times n.$$

Eigenvalues of A:

$$\lambda_{ij} = t_i + t_j - 60, \quad i,j = 1, 2, \ldots, n, \text{ where}$$

$$t_k = 66 - \left(8 + 2\cos\frac{k\pi}{n+1}\right)^2, \quad k = 1, 2, \ldots, n.$$

Reference: [36, pp. 22-24].

<u>Example 4.18</u>

$$A = \begin{bmatrix} X & Y & & & \\ Y & X & Y & & \\ & \cdots & \cdots & \cdots & \\ & & Y & X & Y \\ & & & Y & X \end{bmatrix}, \quad n^2 \times n^2,$$

where

$$X = \begin{bmatrix} -20 & 4 & & & \\ 4 & -20 & 4 & & \\ & \cdots & \cdots & \cdots & \\ & & 4 & -20 & 4 \\ & & & 4 & -20 \end{bmatrix}, \quad n \times n,$$

and

$$Y = \begin{bmatrix} 4 & 1 & & & \\ 1 & 4 & 1 & & \\ & \cdots & \cdots & \cdots & \\ & & 1 & 4 & 1 \\ & & & 1 & 4 \end{bmatrix}, \quad n \times n.$$

Eigenvalues of A:

$$\lambda_{kj} = (-20 - 8\cos k\theta - 8\cos j\theta + 4\cos k\theta \cos j\theta); \; \theta = \frac{\pi}{n+1} \text{ and}$$

$$k,j = 1, 2, \ldots, n.$$

Reference: [36, pp. 22, 24].

Example 4.19

$$A = \begin{bmatrix} -4I_n & X & & & \\ X & -4I_n & X & & \\ & \cdots & \cdots & \cdots & \\ & & X & -4I_n & X \\ & & & X & -4I_n \end{bmatrix}, \quad n^2 \times n^2,$$

where

$$X = \begin{bmatrix} 0 & 1 & & & \\ 1 & 0 & 1 & & \\ & \cdots & \cdots & \cdots & \\ & & 1 & 0 & 1 \\ & & & 1 & 0 \end{bmatrix}, \quad n \times n.$$

Eigenvalues of A:

$$\lambda_{kj} = -4(1+\cos k\theta \cos j\theta); \quad \theta = \frac{\pi}{n+1}, \text{ and } k,j = 1, 2, \ldots, n.$$

Reference: [36, pp. 22, 24].

Example 4.20

$$\begin{bmatrix} -1 & 2a & 1 & & & & \\ 2a & 0 & 2a & 1 & & & \\ 1 & 2a & 0 & 2a & 1 & & \\ & \cdots & \cdots & \cdots & \cdots & \cdots & \\ & & & 1 & 2a & 0 & 2a \\ & & & & 1 & 2a & -1 \end{bmatrix}, \quad n \times n.$$

Eigenvalues:

$$\lambda_k = \left(a-2 \cos \frac{k\pi}{n+1}\right)^2 - (a^2+2), \quad k = 1, 2, \ldots, n.$$

Reference: [18, p. 31].

Example 4.21

Let p be a prime, $p \geq 5$, and let $n = p - 1$. Define $A = [a_{ij}]$ to be the $n \times n$ matrix such that

$$a_{ij} = \begin{cases} 0, & \text{if } p \mid (i+j) \\ 1, & \text{if } i + j \text{ is congruent to a square mod } p \\ -1, & \text{otherwise} \end{cases}$$

Eigenvalues:

$$\lambda_1 = 1$$

$$\lambda_2 = -1$$

$$\lambda_i = \sqrt{p}, \quad i = 3, 4, \ldots, \frac{n}{2} + 1$$

$$\lambda_i = -\sqrt{p} \quad i = \frac{n}{2} + 2, \ldots, n.$$

Inverse: Example 3.14.

Reference: [43].

Example 4.22

Let $A = [a_{ij}]$ be the $n \times n$ matrix defined by

$$a_{ij} = |i-j|.$$

A has a dominant positive eigenvalue and $n-1$ real negative eigenvalues. If $n \equiv 2 \pmod 4$, then $\lambda = -1$ is an eigenvalue with corresponding eigenvector

$$x = \begin{bmatrix} 1 \\ -1 \\ -1 \\ 1 \\ \vdots \end{bmatrix},$$

where the four components shown are repeated periodically.

Inverse: Example 3.23

Reference: [77, pp. 32-33].

Example 4.23 (see Chapter 2, Section 4)

Let J_{nn} be the $n \times n$ matrix all of whose elements are 1. Let f_n denote the column vector all of whose components are 1. Thus

$$J_{nn} = \begin{bmatrix} 1 & 1 & 1 & \dots & 1 \\ 1 & 1 & 1 & \dots & 1 \\ 1 & 1 & 1 & \dots & 1 \\ \hline \\ 1 & 1 & 1 & \dots & 1 \end{bmatrix}, \quad \text{and} \quad f_n = \begin{bmatrix} 1 \\ 1 \\ 1 \\ \vdots \\ 1 \end{bmatrix}.$$

$\lambda_n = n$ is a simple eigenvalue and f_n is its corresponding eigenvector. $\lambda_1 = \lambda_2 = \dots = \lambda_{n-1} = 0$ and every vector orthogonal to f_n is an eigenvector corresponding to $\lambda = 0$. In fact, the $n-1$ dimensional subspace of eigenvectors corresponding to $\lambda = 0$ is spanned by

$$g_i = f_n - n e_n^i, \qquad i = 1, 2, \dots, n-1$$

where e_n^i is the column vector whose components are δ_{ij}, $j = 1, 2, \dots, n$.

Reference: [7], [77].

CHAPTER V

TEST MATRICES: EIGENVALUES AND EIGENVECTORS OF REAL NONSYMMETRIC MATRICES

Example 5.1

$$\begin{bmatrix} 33 & 16 & 72 \\ -24 & -10 & -57 \\ -8 & -4 & -17 \end{bmatrix}$$

Eigenvalues:

$$\lambda_1 = 1$$
$$\lambda_2 = 2$$
$$\lambda_3 = 3$$

Right Eigenvectors:

$$x_1 = \begin{bmatrix} -15 \\ 12 \\ 4 \end{bmatrix}, \qquad x_2 = \begin{bmatrix} -16 \\ 13 \\ 4 \end{bmatrix}, \qquad x_3 = \begin{bmatrix} -4 \\ 3 \\ 1 \end{bmatrix}.$$

Left Eigenvectors:

$$y_1 = [1, 0, 4]$$
$$y_2 = [0, 1, -3]$$
$$y_3 = [4, 4, 3]$$

Inverse: Example 3.1

Reference: [29, pp. 65-69].

Example 5.2

$$\begin{bmatrix} 4 & 1 & 1 \\ 2 & 4 & 1 \\ 0 & 1 & 4 \end{bmatrix}$$

Eigenvalues:

$$\lambda_1 = 3$$
$$\lambda_2 = 3$$
$$\lambda_3 = 6$$

Right Eigenvectors:

$$x_1 = \begin{bmatrix} 0 \\ 1 \\ -1 \end{bmatrix}, \qquad x_3 = \begin{bmatrix} 3 \\ 4 \\ 2 \end{bmatrix}.$$

Left Eigenvectors:

$$y_1 = [2, -1, -1]$$
$$y_3 = [1, 1, 1]$$

Note: Corresponding to the multiple eigenvalue $\lambda_1 = \lambda_2$, we have only one-dimensional subspaces of right and left eigenvectors since the matrix is defective. x_1 and y_1 are vectors from these subspaces.

Reference: [54].

Example 5.3

$$\begin{bmatrix} 1 & 0 & 0.01 \\ 0.1 & 1 & 0 \\ 0 & 1 & 1 \end{bmatrix}$$

Eigenvalues:

$\lambda = 1 + 0.1\omega$ where ω is a cube root of unity, i.e.,

$$\omega \in \left\{1, -\frac{1}{2}\left(1 \pm i\sqrt{3}\right)\right\}$$

Right Eigenvectors:

$$x_\omega = \begin{bmatrix} 1 \\ \omega^2 \\ 10\omega \end{bmatrix}$$

Left Eigenvectors:

$$y_\omega = [1, \omega, 0.1\omega^2]$$

Reference: [17].

Example 5.4.

$$\begin{bmatrix} 8 & -1 & -5 \\ -4 & 4 & -2 \\ 18 & -5 & -7 \end{bmatrix}$$

Tridiagonal Form from Lanczos' Method:

α_i	β_i
8	...
$-\frac{291}{43}$	-86
$\frac{162}{43}$	$\frac{23120}{1849}$

Eigenvalues:

$$\lambda_1 = 2 + 4i$$

$$\lambda_2 = 2 - 4i$$

$$\lambda_3 = 1$$

Right Eigenvectors:

$$x_i = \begin{bmatrix} 1-i \\ 2 \\ -2i \end{bmatrix}, \quad x_2 = \begin{bmatrix} 1+i \\ 2 \\ 2i \end{bmatrix}, \quad x_3 = \begin{bmatrix} 1 \\ 2 \\ 1 \end{bmatrix}.$$

Left Eigenvectors:

$$y_1 = [10, -3-i, -4+2i]$$

$$y_2 = [10, -3+i, -4-2i]$$

$$y_3 = [2, -1, -1]$$

Reference: [22, pp. 256-257].

Example 5.5

$$\begin{bmatrix} -2 & 2 & 2 & 2 \\ -3 & 3 & 2 & 2 \\ -2 & 0 & 4 & 2 \\ -1 & 0 & 0 & 5 \end{bmatrix}$$

Eigenvalues:

$$\lambda_1 = 1$$
$$\lambda_2 = 2$$
$$\lambda_3 = 3$$
$$\lambda_4 = 4$$

Right Eigenvectors:

$$x_1 = \begin{bmatrix} 4 \\ 3 \\ 2 \\ 1 \end{bmatrix}, \quad x_2 = \begin{bmatrix} 3 \\ 3 \\ 2 \\ 1 \end{bmatrix}, \quad x_3 = \begin{bmatrix} 2 \\ 2 \\ 2 \\ 1 \end{bmatrix}, \quad x_4 = \begin{bmatrix} 1 \\ 1 \\ 1 \\ 1 \end{bmatrix}.$$

Left Eigenvectors:

$$y_1 = [1, -1, 0, 0]$$
$$y_2 = [-1, 2, -1, 0]$$
$$y_3 = [0, -1, 2, -1]$$
$$y_4 = [0, 0, -1, 2]$$

Reference: [15].

Example 5.6

$$\begin{bmatrix} 6 & -3 & 4 & 1 \\ 4 & 2 & 4 & 0 \\ 4 & -2 & 3 & 1 \\ 4 & 2 & 3 & 1 \end{bmatrix}$$

Eigenvalues:

$$\lambda_1 = 3 + \sqrt{5}$$

$$\lambda_2 = 3 + \sqrt{5}$$

$$\lambda_3 = 3 - \sqrt{5}$$

$$\lambda_4 = 3 - \sqrt{5}$$

Right Eigenvectors:

$$x_1 = \begin{bmatrix} \sqrt{5} \\ 3 + \sqrt{5} \\ 2 \\ 6 \end{bmatrix}, \quad x_3 = \begin{bmatrix} -\sqrt{5} \\ 3 - \sqrt{5} \\ 2 \\ 6 \end{bmatrix}.$$

Left Eigenvectors:

$$y_1 = [5+\sqrt{5}, -(5+\sqrt{5}), 3\sqrt{5}/2, 5/2]$$

$$y_3 = [5-\sqrt{5}, -(5-\sqrt{5}), -3\sqrt{5}/2, 5/2]$$

Note: Corresponding to the multiple eigenvalue $\lambda_1 = \lambda_2$, we have only one-dimensional subspaces of right and left eigenvectors since the matrix is defective. x_1 and y_1 are vectors from these subspaces. Similarly, x_3 and y_3 are vectors from the one-dimensional subspaces of eigenvectors corresponding to the multiple eigenvalue $\lambda_3 = \lambda_4$.

Reference: [17].

Example 5.7

$$\begin{bmatrix} 0 & 0.07 & 0.27 & -0.33 \\ 1.31 & -0.36 & 1.21 & 0.41 \\ 1.06 & 2.86 & 1.49 & -1.34 \\ -2.64 & -1.84 & -0.24 & -2.01 \end{bmatrix}$$

Eigenvalues:

$$\lambda_1 = 0.03$$

$$\lambda_2 = 3.03$$

$$\lambda_3 = -1.97 + i$$

$$\lambda_4 = -1.97 - i$$

References: [21].

Example 5.8

$$\begin{bmatrix} 4 & -5 & 0 & 3 \\ 0 & 4 & -3 & -5 \\ 5 & -3 & 4 & 0 \\ 3 & 0 & 5 & 4 \end{bmatrix}$$

Eigenvalues:

$$\lambda_1 = 12$$

$$\lambda_2 = 1 + 5i$$

$$\lambda_3 = 1 - 5i$$

$$\lambda_4 = 2$$

Right Eigenvectors:

$$x_1 = \begin{bmatrix} 1 \\ -1 \\ 1 \\ 1 \end{bmatrix}, \quad x_2 = \begin{bmatrix} 1 \\ -i \\ -i \\ -1 \end{bmatrix}, \quad x_3 = \begin{bmatrix} 1 \\ i \\ i \\ -1 \end{bmatrix}, \quad x_4 = \begin{bmatrix} 1 \\ 1 \\ -1 \\ 1 \end{bmatrix}.$$

Left Eigenvectors:

$$y_1 = [1, -1, 1, 1]$$

$$y_2 = [1, i, i, -1]$$

$$y_3 = [1, -i, -i, -1]$$

$$y_4 = [1, 1, -1, 1]$$

Reference: [49, pp. 57-58], [60, p. 147].

Example 5.9

$$\begin{bmatrix} 122 & 41 & 40 & 26 & 25 \\ 40 & 170 & 25 & 14 & 24 \\ 27 & 26 & 172 & 7 & 3 \\ 32 & 22 & 9 & 106 & 6 \\ 31 & 28 & -2 & -1 & 165 \end{bmatrix}$$

Eigenvalues:

$$\lambda_1 \doteq 242.97727\ 3320$$

$$\lambda_2 \doteq 167.48487\ 8917$$

$$\lambda_3 \doteq 134.68646\ 3320$$

$$\lambda_4 \doteq 112.15419\ 3247$$

$$\lambda_5 \doteq 77.69719\ 11963$$

Reference: [38].

Example 5.10

$$\begin{bmatrix} 0.4163 & 0.3176 & 0 & 0 & 0 \\ 0.0001 & 0.4132 & 0.8175 & 0 & 0 \\ 0.6321 & 0.3157 & 0.4823 & 0.6614 & 0 \\ 0.5157 & 0.8321 & 0.5642 & 0.6541 & 0.4321 \\ 0.5563 & 0.4431 & 0.2567 & 0.8325 & 0.8475 \end{bmatrix}$$

Tridiagonal Form from the Elimination Method:

$$\begin{bmatrix} 0.4163 & 0.3176 & 0 & 0 & 0 \\ 0.0001 & 5167.8307 & 0.8175 & 0 & 0 \\ 0 & -3265\ 9398.0809 & -5166.3956\ 7804 & 0.6614 & 0 \\ 0 & 0 & -0.0001\ 1909 & 0.5615 & 0.4321 \\ 0 & 0 & 0 & 0.5463 & 0.4006 \end{bmatrix}$$

Eigenvalues:

$$\lambda_1 \doteq 1.8390$$

$$\lambda_2 \doteq 0.2363$$

$$\lambda_3 \doteq 0.8045$$

$$\lambda_4 \doteq -0.0332 + 0.4374i$$

$$\lambda_5 \doteq -0.0332 - 0.4374i$$

Reference: [69].

Example 5.11

$$\begin{bmatrix} 15 & 11 & 6 & -9 & -15 \\ 1 & 3 & 9 & -3 & -8 \\ 7 & 6 & 6 & -3 & -11 \\ 7 & 7 & 5 & -3 & -11 \\ 17 & 12 & 5 & -10 & -16 \end{bmatrix}$$

Characteristic Polynomial:

$$P(\lambda) = \lambda^5 - 5\lambda^4 + 33\lambda^3 - 51\lambda^2 + 135\lambda + 225$$

Eigenvalues:

$$\lambda_1 = 1.5 + \sqrt{12.75}\, i$$
$$\lambda_2 = 1.5 + \sqrt{12.75}\, i$$
$$\lambda_3 = 1.5 - \sqrt{12.75}\, i$$
$$\lambda_4 = 1.5 - \sqrt{12.75}\, i$$
$$\lambda_5 = -1$$

Right Eigenvectors:

$$x_1 = \begin{bmatrix} 184 - 230\sqrt{12.75}\, i \\ 507 - 52\sqrt{12.75}\, i \\ 295.5 - 163\sqrt{12.75}\, i \\ 411 - 166\sqrt{12.75}\, i \\ 213.5 - 223\sqrt{12.75}\, i \end{bmatrix}, \quad x_3 = \begin{bmatrix} 184 + 230\sqrt{12.75}\, i \\ 507 + 52\sqrt{12.75}\, i \\ 295.5 + 163\sqrt{12.75}\, i \\ 411 + 166\sqrt{12.75}\, i \\ 213.5 + 223\sqrt{12.75}\, i \end{bmatrix}, \quad x_5 = \begin{bmatrix} 13 \\ 22 \\ 19 \\ 16 \\ 28 \end{bmatrix}$$

Left Eigenvectors:

$$y_1 = \left[150 - 342\sqrt{12.75}\, i,\ 184 - 230\sqrt{12.75}\, i,\ -1630 - 74\sqrt{12.75}\, i,\ 589.5 + 409\sqrt{12.75}\, i,\ 555 + 156\sqrt{12.75}\, i\right]$$

$$y_3 = \left[150 + 342\sqrt{12.75}\, i,\ 184 + 230\sqrt{12.75}\, i,\ -1630 + 74\sqrt{12.75}\, i,\right.$$

$$\left. 589.5 - 409\sqrt{12.75}\, i,\ 555 - 156\sqrt{12.75}\, i\right]$$

$$y_5 = [-25, -3, 16, -7, 20]$$

Note: Corresponding to the multiple eigenvalue $\lambda_1 = \lambda_2$, we have only one-dimensional subspaces of right and left eigenvectors since the matrix is defective. x_1 and y_2 are vectors from these subspaces. Similarly, x_3 and y_3 are vectors from the one-dimensional subspaces of eigenvectors corresponding to the multiple eigenvalue λ_3 and λ_4.

Reference: [15], [51], [60, pp. 143-144].

Example 5.12

$$\begin{bmatrix} 10 & -19 & 17 & -12 & 4 & 1 \\ 9 & -18 & 17 & -12 & 4 & 1 \\ 8 & -16 & 15 & -11 & 4 & 1 \\ 6 & -12 & 12 & -10 & 4 & 1 \\ 4 & -8 & 8 & -6 & 1 & 2 \\ 2 & -4 & 4 & -3 & 1 & 0 \end{bmatrix}$$

Eigenvalues:

$$\lambda_1 = -1$$
$$\lambda_2 = -1$$
$$\lambda_3 = -1$$
$$\lambda_4 = 1$$
$$\lambda_5 = i$$
$$\lambda_6 = -i$$

Reference: [58].

Example 5.13

$$A = \begin{bmatrix} B & 2B \\ 4B & 3B \end{bmatrix}$$

where

$$B = \begin{bmatrix} 0 & 1 & 0 & 0 & 0 \\ 0 & 0 & 1 & 0 & 0 \\ 0 & 0 & 0 & 1 & 0 \\ 0 & 0 & 0 & 0 & 1 \\ \epsilon & 0 & 0 & 0 & 0 \end{bmatrix}$$

with $\epsilon = 10^{-5}$.

Eigenvalues of A:

$$\lambda_k = 0.5 \exp(2k\pi i/5), \quad k = 1, 2, 3, 4, 5.$$

$$\lambda_k = -0.1 \exp(2k\pi i/5), \quad k = 1, 2, 3, 4, 5.$$

Reference: [45], [61].

Example 5.14

$$\begin{bmatrix} 12 & 11 & & & & & \\ 11 & 11 & 10 & & & & \\ 10 & 10 & 10 & 9 & & & \\ & \cdots & \cdots & \cdots & \cdots & \cdots & \\ 2 & 2 & 2 & 2 & \cdots & 2 & 1 \\ 1 & 1 & 1 & 1 & \cdots & 1 & 1 \end{bmatrix}$$

Eigenvalues:

$$\lambda_1 \doteq 32.22889\ 15015\ 72160\ 750$$

$$\lambda_2 \doteq 20.19898\ 86458\ 77079\ 428$$

$$\lambda_3 \doteq 12.31107\ 74088\ 68526\ 120$$

$$\lambda_4 \doteq 6.96153\ 30855\ 67122\ 113$$

$$\lambda_5 \doteq 3.51185\ 59485\ 80757\ 194$$

$$\lambda_6 \doteq 1.55398\ 87091\ 32107\ 90$$

$\lambda_7 \doteq 0.64350\ 53190\ 04855\ 5$

$\lambda_8 \doteq 0.28474\ 97205\ 58478$

$\lambda_9 \doteq 0.14364\ 65197\ 69220$

$\lambda_{10} \doteq 0.08122\ 76592\ 40405$

$\lambda_{11} \doteq 0.04950\ 74291\ 85278$

$\lambda_{12} \doteq 0.03102\ 80606\ 44010$

We give the right and left eigenvectors corresponding to λ_{10}, λ_{11}, λ_{12}, the three most sensitive eigenvalues. See Varah [79, pp. 107-111] for additional vectors.

Right Eigenvectors:

x_{10}	x_{11}	x_{12}
-0.67714 11	-0.67599 46	-0.67531 89
0.73369 91	0.73440 62	0.73480 66
-0.05625 42	-0.06061 69	-0.06315 65
-0.00084 53	0.00211 69	0.00385 87
0.00060 06	0.00011 26	-0.00019 87
-0.00006 06	-0.00002 68	0.00000 91
0.00000 09	0.00000 29	-0.00000 04
0.00000 07	-0.00000 02	0.00000 00
-0.00000 01	0.00000 00	0.00000 00
0.00000 00	0.00000 00	0.00000 00
0.00000 00	0.00000 00	0.00000 00
0.00000 00	0.00000 00	0.00000 00

Left Eigenvectors:

y_{10}	y_{11}	y_{12}
0.00000 00	0.00000 00	0.00000 00
0.00000 00	0.00000 00	0.00000 00
0.00000 00	0.00000 00	-0.00000 03
-0.00000 10	0.00000 03	0.00000 38
0.00001 10	-0.00001 41	-0.00004 51
-0.00002 37	0.00021 95	0.00043 20
-0.00068 17	-0.00221 72	-0.00331 87
0.00946 06	0.01596 10	0.02009 76
-0.06504 47	-0.08251 35	-0.09284 49
0.26984 68	0.29459 26	0.30854 36
-0.64994 86	-0.65581 13	-0.65861 25
0.70740 99	0.68997 00	0.67970 23

Condition Numbers:

$s_1 \doteq 0.30424\ 083$	$s_7 \doteq 0.00006\ 913$
$s_2 \doteq -0.20079\ 033$	$s_8 \doteq -0.00000\ 178$
$s_3 \doteq 0.31822\ 599$	$s_9 \doteq 0.00000\ 01498$
$s_4 \doteq -0.58447\ 355$	$s_{10} \doteq -0.00000\ 00375$
$s_5 \doteq 0.14446\ 703$	$s_{11} \doteq 0.00000\ 00258$
$s_6 \doteq -0.00462\ 656$	$s_{12} \doteq -0.00000\ 00547$

References: [17], [71], [72, pp. 151-153], [79, pp. 106-111].

Example 5.15

38747	26624	46397	21041	10983
-49239	523895	874986	518535	207320
-125005	760217	1889996	987939	812227
-175215	707955	1585904	1835031	1452154
-176459	356979	1421590	1490020	2432634
-68786	13606	-1635	5801	1970
-70392	74578	-20029	32720	11970
-66818	36906	-35373	66689	28893 ...
-56793	-7997	-190532	121047	65580
-39309	-47171	-336666	-92801	165443
-15085	-76321	-459117	-276532	-166380
-107826	-54142	-47535	-18910	-4472
-30063	-15101	-13276	-5304	-1304
-10446	-5252	-4631	-1871	-501
-31693	-15937	-14052	-5675	-1523

-74761	-51553	-49730	-39456	-41365
-9747	94918	41968	-15257	-67307
-65257	-20436	-30683	-157418	-305659
-59161	78513	149178	188162	-110555
-60822	62926	121436	152216	225707
162836	110162	101956	76674	75926
147007	696799	645055	485251	480500
... 135534	690486	1247127	938227	929431 ...
127430	682860	1233574	1542211	1528044
125487	675886	1220769	1526106	2189518
129710	670239	1209950	1512267	2169459
-12116	-1960	-948	157	1128
-3369	-547	-266	42	311
-1163	-190	-93	12	105
-3527	-578	-284	38	320

	-34389	-16813	-50316	-40743	-31954
	-105605	-182666	-595834	-527558	-439963
	-390172	-363323	-1185925	-1044921	-863211
	-389762	-400208	-1327062	-1183861	-979778
	-285526	-358272	-1202147	-1079850	-890416
	58807	11793	30018	19232	11895
	372486	23216	49257	15034	-6149
...	720265	41433	95189	38606	2197
	1184459	70143	178721	97822	41610
	1698300	108178	298506	194757	115175
	2369651	153437	448574	324414	217277
	1869	1536234	18473	-519690	-438975
	518	285553	108570	-144914	-122917
	177	-18787	-58126	-932	110583
	536	-57006	-176348	12799	317581

Eigenvalues:

$\lambda_1 \doteq 6294127.73$

$\lambda_2 \doteq 4830173.88$

$\lambda_3 \doteq 1593256.12$

$\lambda_4 \doteq 1296443.15$

$\lambda_5 \doteq 976578.96$

$\lambda_6 \doteq 517836.12$

$\lambda_7 \doteq 369921.48$

$\lambda_8 \doteq 308201.60$

$\lambda_9 \doteq 257611.17$

$\lambda_{10} \doteq 173583.27$

$\lambda_{11} \doteq 151487.87$

$\lambda_{12} \doteq 73704.19$

$\lambda_{13} \doteq 43990.81$

$\lambda_{14} \doteq 3587.14$

$\lambda_{15} \doteq -605.47$

Right Eigenvectors:

x_1	x_2	x_3	x_4	x_5
-25600791	8179865	-10559890	-6382980	71337697
-93333619	281884032	-154642407	-514788753	230978290
-281658289	618796420	-135528630	-635147361	463951989
-234339803	892390433	172711194	-51309046	-149941032
-180185122	1000000000	567634488	1000000000	-79550827
41356495	13057924	1504059	-42915249	-111068725
255125017	78558762	-37368379	-228515241	-582184747
466609433	136928986	-47836788	-275319457	-737157698
693438409	179243759	-16468980	-90455002	-393746553
876398341	166202576	18990089	200223576	310698214
1000000000	51289932	-111765187	39001492	1000000000
5717639	-19636264	1000000000	-154945349	38153292
1491503	-5022691	202341288	-28449817	5665174
425148	-1324500	-23478411	10379135	-6683085
1287435	-4008552	-70672023	31184833	-20016569

x_6	x_7	x_8	x_9	x_{10}
35546791	-96033014	-43838567	-125099214	-316813407
18937820	88496756	114088163	590391245	-374763409
-430319318	-381761741	584957693	-214921796	302668034
1000000000	622739632	189896517	85285698	-282248681
-469843941	65080697	42051926	-147661918	117044666
-81892147	195869218	74731927	343989557	605390857
-353035016	662203426	-221204146	777971729	84552451
-211843343	-12939707	17654019	-957697355	-568699623
425924519	-716871720	163015274	-102833479	1000000000
-37494408	-632446333	-16984885	1000000000	-688507740
-30655114	1000000000	-48074780	-706623434	260312429
-17087	15840072	532263383	35799075	-40233177
8560	-2494823	-150431718	-16577034	54361072
340161	15190410	343042433	17464795	-9100840
1016943	44562429	1000000000	50539011	-25757337

x_{11}	x_{12}	x_{13}	x_{14}	x_{15}
-483941289	80100044	158500974	1000000000	926568323
-54509219	1000000000	1000000000	457078895	941981992
-25563085	-683069341	217439196	509694713	817688917
294522536	-560808914	135628883	387188842	609029854
-179052534	319086961	331963034	208187123	382268933
1000000000	329734642	-211818796	187142783	-26357098
-718306625	-326232825	-94574848	10731580	-11586613
560841987	140258619	29608774	-13901773	-18513917
-700735268	307477539	78272228	-29683858	-32489449
509633733	-361747416	-140465927	-14317443	-47897332
-167348227	109768320	-7629423	-38506599	-62329715
-33933888	-88089124	291730040	403471483	686089278
71277033	-367241747	646206560	576215354	947542071
-2855789	-96270976	231004723	435241146	1000000000
-7992436	-251404803	556214353	536274143	795951845

Reference: [73].

Example 5.16

$$A = \begin{bmatrix} B & 2B \\ 4B & 3B \end{bmatrix}$$

where

$$B = \begin{bmatrix} 5C & -C \\ 5C & C \end{bmatrix}$$

and

$$C = \begin{bmatrix} -2 & 2 & 2 & 2 \\ -3 & 3 & 2 & 2 \\ -2 & 0 & 4 & 2 \\ -1 & 0 & 0 & 5 \end{bmatrix}$$

Eigenvalues of A:

$\lambda_1 = 15 + 5\,i$ $\qquad \lambda_9 = -3 + i$

$\lambda_2 = 15 - 5\,i$ $\qquad \lambda_{10} = -3 - i$

$\lambda_3 = 30 + 10\,i$ $\qquad \lambda_{11} = -6 + 2\,i$

$\lambda_4 = 30 - 10\,i$ $\qquad \lambda_{12} = -6 - 2\,i$

$\lambda_5 = 45 + 15\,i$ $\qquad \lambda_{13} = -9 + 3\,i$

$\lambda_6 = 45 - 15\,i$ $\qquad \lambda_{14} = -9 - 3\,i$

$\lambda_7 = 60 + 20\,i$ $\qquad \lambda_{15} = -12 + 4\,i$

$\lambda_8 = 60 - 20\,i$ $\qquad \lambda_{16} = -12 - 4\,i$

Reference: [15].

Example 5.17

$$A = \begin{bmatrix} 8B & 4B \\ -5B & -B \end{bmatrix},$$

where

$$B = \begin{bmatrix} C & 2C \\ 4C & 3C \end{bmatrix},$$

$$C = \begin{bmatrix} 4D & 3D \\ -2D & -D \end{bmatrix},$$

$$D = \begin{bmatrix} -1 & 1 & 0 & 0 & 0 \\ -1 & 0 & 1 & 0 & 0 \\ -1 & 0 & 0 & 1 & 0 \\ -1 & 0 & 0 & 0 & 1 \\ -1 & 0 & 0 & 0 & 0 \end{bmatrix}.$$

Eigenvalues of A:

$8, 6, 4, 3, -40, -30, -20, -15, \pm 40j, \pm 30j, \pm 20j, \pm 15j,$

$\pm 8j, \pm 6j, \pm 4j, \pm 3j$ where $j = \frac{1}{2}\left(1 \pm \sqrt{3}\, i\right).$

Reference: [15], [45], [61].

Example 5.18

$$A = \begin{bmatrix} B & 2B \\ 4B & 3B \end{bmatrix}$$

where

$$B = \begin{bmatrix} 3C & 3C \\ 5C & C \end{bmatrix},$$

$$C = \begin{bmatrix} 6D & -D & D & 0 \\ 8D & 0 & D & 2D \\ -2D & 0 & D & 2D \\ 5D & -D & -D & D \end{bmatrix},$$

$$D = \begin{bmatrix} -2 & 2 & 2 & 2 \\ -3 & 3 & 2 & 2 \\ -2 & 0 & 4 & 2 \\ -1 & 0 & 0 & 5 \end{bmatrix} .$$

Eigenvalues of A:

120j, 90j, 60j, 30j, -40j, -30j, -20j, -10j, -24j, -18j,

-12j, -6j, 8j, 6j, 4j, 2j where $j \in \{3 \pm i, 1 \pm 2i\}$.

Reference: [45].

Example 5.19

Let $A = [a_{ij}]$ be the 100 x 100 matrix defined by

$$a_{ij} = \begin{cases} 101 - i, & \text{if } j = i \\ (-1)^{i+j+1} \, 40/(i+j-2), & \text{if } j < i \\ 40/102, & \text{if } i = 1 \text{ and } j = 2 \\ 40, & \text{if } i = 1 \text{ and } j = 100 \\ 0, & \text{otherwise.} \end{cases}$$

Eigenvalues of A:

1, 2, 3, ..., 99, 100.

Reference: [45].

Example 5.20

Let $A_n = [a_{ij}]$ be the n x n matrix defined by

$a_{1j} = 1, j = 1, 2, \ldots, n,$

$a_{ij} = (i+j-1)^{-1}, i = 2, 3, \ldots, n, j = 1, 2, \ldots, n.$

$$A_n = \begin{bmatrix} 1 & 1 & 1 & \cdots & 1 \\ \frac{1}{2} & \frac{1}{3} & \frac{1}{4} & \cdots & \frac{1}{n+1} \\ \frac{1}{3} & \frac{1}{4} & \frac{1}{5} & \cdots & \frac{1}{n+2} \\ \cdots & \cdots & \cdots & \cdots & \cdots \\ \frac{1}{n} & \frac{1}{n+1} & \frac{1}{n+2} & \cdots & \frac{1}{2n-1} \end{bmatrix}$$

Eigenvalues:

Let $\lambda_M(n)$ and $\lambda_m(n)$ be the eigenvalues of A_n of largest and smallest magnitude respectively.

n	$\lambda_M(n)$	$\lambda_m(n)$
2	1.448 403	-0.115 0693
3	1.707 105	$-0.481\ 5399 \times 10^{-2}$
4	1.886 632	$-0.144\ 1324 \times 10^{-3}$
5	2.022 999	$-0.448\ 9833 \times 10^{-5}$
6	2.132 376	$-0.139\ 7499 \times 10^{-6}$
7	2.223 362	$-0.433\ 6577 \times 10^{-8}$
8	2.301 055	$-0.134\ 0623 \times 10^{-9}$
9	2.368 717	$-0.412\ 9309 \times 10^{-11}$
10	2.428 554	$-0.126\ 7649 \times 10^{-12}$

Right eigenvector corresponding to $\lambda_M(n)$:

n = 2	3	4	5	6
1.000000	1.000000	1.000000	1.000000	1.000000
0.448403	0.416793	0.394224	0.376917	0.363036
	0.290313	0.277320	0.266965	0.258417
		0.215088	0.208099	0.202206
			0.171019	0.166679
				0.142038

7	8	9	10
1.000000	1.000000	1.000000	1.000000
0.351545	0.341809	0.333408	0.326051
0.251183	0.244943	0.239479	0.234636
0.197135	0.192700	0.188772	0.185257
0.162893	0.159546	0.156554	0.153855
0.139091	0.136461	0.134092	0.131940
0.121514	0.119387	0.117457	0.115694
	0.106209	0.104603	0.103129
		0.094352	0.093099
			0.084893

Right eigenvector corresponding to $\lambda_m(n)$:

n = 2	3	4	5	6
-0.896805	0.550163	-0.224188	0.081470	-0.027443
1.000000	-1.552812	1.270271	-0.772313	0.392118
	1.000000	-2.046050	2.237419	-1.768987
		1.000000	-2.546576	3.450596
			1.000000	-3.046283
				1.000000

7	8	9	10
0.008795	-0.002721	0.000819	-0.000242
-0.176722	0.073168	-0.028430	0.010516
1.140790	-0.638751	0.322286	-0.150100
-3.339147	2.598579	-1.732159	1.027660
4.912246	-5.607667	5.099020	-3.934520
-3.545962	6.623060	-8.699495	9.038123
1.000000	-4.045669	8.583374	-12.739602
	1.000000	-4.545416	10.793366
		1.000000	-5.045200
			1.000000

Eigenvalues of A_6:

$$\lambda_1 \doteq 2.1323\ 763$$

$$\lambda_2 \doteq -0.2214\ 0681$$

$$\lambda_3 \doteq -0.3184\ 3305 \times 10^{-1}$$

$$\lambda_4 \doteq -0.8983\ 2330 \times 10^{-3}$$

$$\lambda_5 \doteq -0.1706\ 2788 \times 10^{-4}$$

$$\lambda_6 \doteq -0.1397\ 4990 \times 10^{-6}$$

Inverse: Example 3.9

Reference: [32], [33].

Example 5.21

$$A = \begin{bmatrix} -1 & -1 & -1 & \cdots & -1 & -1 \\ 1 & 0 & 0 & \cdots & 0 & 0 \\ 0 & 1 & 0 & \cdots & 0 & 0 \\ & \cdots & \cdots & \cdots & \cdots & \\ 0 & 0 & 0 & \cdots & 1 & 0 \end{bmatrix}, \quad n \times n.$$

Eigenvalues:

$$\lambda_k = \exp \frac{2k\pi i}{n+1}, \quad k = 1, 2, \ldots, n.$$

Right Eigenvectors:

Let $x^{(k)}$ be the right eigenvector of A corresponding to the eigenvalue λ_k, $k = 1, 2, \ldots, n$. Then

$$x^{(k)} = \begin{bmatrix} \lambda_k^{n-1} \\ \lambda_k^{n-2} \\ \vdots \\ \lambda_k \\ 1 \end{bmatrix}$$

Left Eigenvectors:

Let $y^{(k)} = \left[y_1^{(k)}, y_2^{(k)}, \ldots, y_n^{(k)}\right]$ be the left eigenvector of A corresponding to the eigenvalue λ_k, $k = 1, 2, \ldots, n$. Then, for each k,

$$y_j^{(k)} = \frac{\sum_{m=0}^{n-j} \lambda_k^m}{\lambda_k^{n-j}}, \quad j = 1, 2, \ldots, n.$$

Reference: [17].

Example 5.22 (Forsythe)

$$\begin{bmatrix} 0 & 1 & & & & & \\ & 0 & 1 & & & & \\ & & 0 & 1 & & & \\ & & & \cdot & \cdot & & \\ & & & & \cdot & \cdot & \\ & & & & & \cdot & \cdot \\ & & & & & 0 & 1 \\ \epsilon & & & & & & 0 \end{bmatrix}$$

Characteristic Equation:

$$\lambda^n - \epsilon = 0$$

Eigenvalues:

$$\lambda_k = \sqrt[n]{|\epsilon|} \exp \frac{2k\pi i}{n}, \quad k = 1, 2, \ldots, n.$$

Reference: [61], [76], [62, p. 64].

Example 5.23

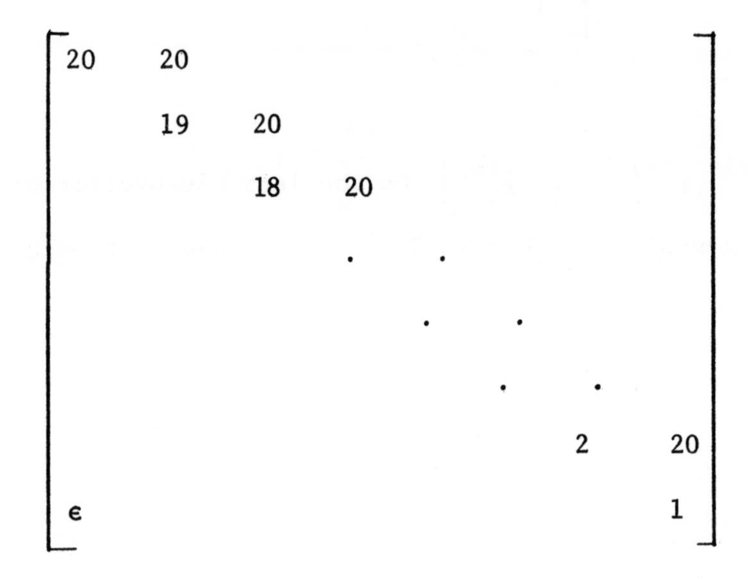

Characteristic Equation:

$$(20-\lambda)(19-\lambda) \ldots (1-\lambda) - 20^{19}\epsilon = 0$$

Eigenvalues:

If $\epsilon = 10^{-10}$, the eigenvalues are

0.99575439	3.96533070 ± 1.08773570i
20.00424561	17.03466930 ± 1.08773570i
2.10924184	5.89397755 ± 1.94852927i
18.89075816	15.10602245 ± 1.94852927i
2.57488140	8.11807338 ± 2.52918173i
18.42511860	12.88192662 ± 2.52918173i
	10.50000000 ± 2.73339736i

Condition Numbers:

If $\epsilon = 0$, $s_k = \frac{(20-k)!(k-1)!}{20^{19}}$, $k = 1, 2, \ldots, 20$, is the condition number corresponding to the eigenvalue $\lambda = k$.

Reference: [62, pp. 90-91].

Example 5.24

For arbitrary constants $a_1, a_2, \ldots, a_{n-1}$, let

$$A_n = A_n(a_1,a_2,\ldots,a_{n-1})$$
$$= [a_{ij}]$$

denote the $n \times n$ matrix defined by

$$a_{ij} = \begin{cases} 1 & j \geq i, \\ a_j & j < i. \end{cases}$$

For example,

$$A_4 = \begin{bmatrix} 1 & 1 & 1 & 1 \\ a_1 & 1 & 1 & 1 \\ a_1 & a_2 & 1 & 1 \\ a_1 & a_2 & a_3 & 1 \end{bmatrix}.$$

If we define

$$P_j(\lambda) = (\lambda-1+a_1)(\lambda-1+a_2) \ldots (\lambda-1+a_j), \quad j = 1, 2, \ldots, n-1,$$

we can write the characteristic polynomial of A_n as

$$\lambda^n - \lambda^{n-1} - \lambda^{n-2}P_1(\lambda) - \lambda^{n-3}P_2(\lambda) - \ldots - P_{n-1}(\lambda).$$

If $\lambda \neq 0$ is an eigenvalue of A_n and if x is a right eigenvector corresponding to λ, it can be shown that

$$x = \begin{bmatrix} 1 \\ x_2 \\ x_3 \\ \vdots \\ x_n \end{bmatrix}$$

where

$$x_j = \lambda^{-(j-1)} P_{j-1}(\lambda), \qquad j = 2, 3, \ldots, n.$$

If $\lambda \neq 0$ is a multiple eigenvalue of A, there is only a one-dimensional subspace of eigenvectors associated with λ, i.e., A_n is defective.

If $\lambda = 0$ is an eigenvalue of A_n, the expression for $\det(A_n)$ shows that $a_i = 1$ for certain values of i. Suppose that $\lambda = 0$ is an eigenvalue of multiplicity k and that $a_{i_1} = \ldots = a_{i_k} = 1$. Then there is a k-dimensional subspace of eigenvectors associated with $\lambda = 0$, and the vectors

$$e_n - e_{i_j}, \quad j = 1, 2, \ldots, k,$$

form a basis for the subspace.

Inverse: Example 3.24.

Reference: [75].

Example 5.25

$$A = \begin{bmatrix} -\frac{1}{3} & \frac{1}{6} & 0 & \frac{9}{2} & -3 & -1 \\ -\frac{4}{3} & \frac{2}{3} & 0 & 9 & -6 & -2 \\ -\frac{4}{3} & -\frac{5}{6} & 1 & \frac{27}{2} & -9 & -3 \\ -\frac{4}{3} & -\frac{5}{6} & -1 & \frac{39}{2} & -12 & -4 \\ -\frac{4}{3} & -\frac{5}{6} & -1 & \frac{43}{2} & -13 & -5 \\ -\frac{4}{3} & -\frac{5}{6} & -1 & \frac{43}{2} & -10 & -8 \end{bmatrix}$$

Eigenvalues:

$$\lambda_1 = -3$$

$$\lambda_2 = -2$$

$$\lambda_3 = 2$$

$$\lambda_4 = \frac{3}{2}$$

$$\lambda_5 = 1$$

$$\lambda_6 = \frac{1}{3}$$

Right Eigenvectors:

$$x_1 = \frac{1}{6}\begin{bmatrix} 1 \\ 2 \\ 3 \\ 4 \\ 5 \\ 6 \end{bmatrix}, \quad x_2 = \frac{1}{5}\begin{bmatrix} 1 \\ 2 \\ 3 \\ 4 \\ 5 \\ 5 \end{bmatrix}, \quad x_3 = \frac{1}{4}\begin{bmatrix} 1 \\ 2 \\ 3 \\ 4 \\ 4 \\ 4 \end{bmatrix},$$

$$x_4 = \frac{1}{3}\begin{bmatrix} 1 \\ 2 \\ 3 \\ 3 \\ 3 \\ 3 \end{bmatrix}, \quad x_5 = \frac{1}{2}\begin{bmatrix} 1 \\ 2 \\ 2 \\ 2 \\ 2 \\ 2 \end{bmatrix}, \quad x_6 = \begin{bmatrix} 1 \\ 1 \\ 1 \\ 1 \\ 1 \\ 1 \end{bmatrix}$$

Reference: [79, p. 100].

Example 5.26

$$A = \begin{bmatrix} -9 & 21 & -15 & 4 & 2 & 0 \\ -10 & 21 & -14 & 4 & 2 & 0 \\ -8 & 16 & -11 & 4 & 2 & 0 \\ -6 & 12 & -9 & 3 & 3 & 0 \\ -4 & 8 & -6 & 0 & 5 & 0 \\ -2 & 4 & -3 & 0 & 1 & 3 \end{bmatrix}$$

This matrix is defective and also derogatory [80]. It has an eigenvalue of multiplicity 2 corresponding to a quadratic elementary divisor and an eigenvalue of multiplicity 2 corresponding to two (equal) linear elementary divisors. The other two eigenvalues are complex conjugates.

Eigenvalues:

$$\lambda_1 = \lambda_2 = 3 \qquad \text{(linear elementary divisors)}$$

$$\lambda_3 = 2 + i$$

$$\lambda_4 = 2 - i$$

$$\lambda_5 = \lambda_6 = 1 \qquad \text{(a quadratic elementary divisor)}$$

Eigenvectors:

Corresponding to $\lambda_1 = \lambda_2$ we have the two-dimensional subspace of eigenvectors spanned by

$$x_1 = \begin{bmatrix} 1 \\ 1 \\ 1 \\ 1 \\ 1 \\ 0 \end{bmatrix}, \qquad x_2 = \begin{bmatrix} 0 \\ 0 \\ 0 \\ 0 \\ 0 \\ 1 \end{bmatrix}.$$

Corresponding to λ_3 and λ_4 we have x_3 and x_4 given by

$$\frac{1}{61} \begin{bmatrix} 61 \\ 5(11 \pm i) \\ 4(11 \pm i) \\ 3(11 \pm i) \\ 2(11 \pm i) \\ (11 \pm i) \end{bmatrix}.$$

Corresponding to $\lambda_5 = \lambda_6$ we have the <u>eigenvector</u> x_5 and the <u>principal vector</u> of degree two, x_6, given by

$$x_5 = \frac{1}{4} \begin{bmatrix} 4 \\ 4 \\ 4 \\ 3 \\ 2 \\ 1 \end{bmatrix}, \qquad x_6 = \frac{1}{3} \begin{bmatrix} 3 \\ 3 \\ 3 \\ 3 \\ 2 \\ 1 \end{bmatrix}.$$

These six normalized vectors (five eigenvectors and a principal vector) make up the transformation to Jordan canonical form.

Reference: [79, p. 103 and p. 207].

Example 5.27

$$A = \begin{bmatrix} 1 & 1 & 1 & -2 & 1 & -1 & 2 & -2 & 4 & -3 \\ -1 & 2 & 3 & -4 & 2 & -2 & 4 & -4 & 8 & -6 \\ -1 & 0 & 5 & -5 & 3 & -3 & 6 & -6 & 12 & -9 \\ -1 & 0 & 3 & -4 & 4 & -4 & 8 & -8 & 16 & -12 \\ -1 & 0 & 3 & -6 & 5 & -4 & 10 & -10 & 20 & -15 \\ -1 & 0 & 3 & -6 & 2 & -2 & 12 & -12 & 24 & -18 \\ -1 & 0 & 3 & -6 & 2 & -5 & 15 & -13 & 28 & -21 \\ -1 & 0 & 3 & -6 & 2 & -5 & 12 & -11 & 32 & -24 \\ -1 & 0 & 3 & -6 & 2 & -5 & 12 & -14 & 37 & -26 \\ -1 & 0 & 3 & -6 & 2 & -5 & 12 & -14 & 36 & -25 \end{bmatrix}$$

This matrix is both defective and derogatory [80]. As a matter of fact, a multiple eigenvalue is associated with more than one nonlinear elementary divisor in two instances. To be specific, $\lambda = 2$ is an eigenvalue of multiplicity 5 but it is associated with two nonlinear elementary divisors, one of degree 3 and one of degree 2. Likewise, $\lambda = 3$ is an eigenvalue of multiplicity 4 but it is associated with two quadratic elementary divisors. Consequently, the results are grouped according to the invariant subspaces spanned by the eigenvectors and principal vectors shown.

Eigenvalues, Eigenvectors, and Principal Vectors:

$\lambda_1 = \lambda_2 = \lambda_3 = 2$ is associated with the eigenvector x_1 and the principal vectors x_2 and x_3 of degrees 2 and 3, respectively, given by

$$x_1 = \begin{bmatrix} 1 \\ 1 \\ 1 \\ 1 \\ 1 \\ 1 \\ 1 \\ 1 \\ 1 \\ 1 \end{bmatrix}, \quad x_2 = \frac{1}{2}\begin{bmatrix} 1 \\ 2 \\ 2 \\ 2 \\ 2 \\ 2 \\ 2 \\ 2 \\ 2 \\ 2 \end{bmatrix}, \quad x_3 = \frac{1}{3}\begin{bmatrix} 1 \\ 2 \\ 3 \\ 3 \\ 3 \\ 3 \\ 3 \\ 3 \\ 3 \\ 3 \end{bmatrix}.$$

$\lambda_4 = \lambda_5 = 2$ is associated with the <u>eigenvector</u> x_4 and the <u>principal vector</u> x_5 of degree 2 given by

$$x_4 = \frac{1}{4}\begin{bmatrix} 1 \\ 2 \\ 3 \\ 4 \\ 4 \\ 4 \\ 4 \\ 4 \\ 4 \\ 4 \end{bmatrix}, \quad x_5 = \frac{1}{5}\begin{bmatrix} 1 \\ 2 \\ 3 \\ 4 \\ 5 \\ 5 \\ 5 \\ 5 \\ 5 \\ 5 \end{bmatrix}.$$

$\lambda_6 = \lambda_7 = 3$ is associated with the <u>eigenvector</u> x_6 and the <u>principal vector</u> x_7 of degree 2 given by

$$x_6 = \frac{1}{6}\begin{bmatrix} 1 \\ 2 \\ 3 \\ 4 \\ 5 \\ 6 \\ 6 \\ 6 \\ 6 \\ 6 \end{bmatrix}, \qquad x_7 = \frac{1}{7}\begin{bmatrix} 1 \\ 2 \\ 3 \\ 4 \\ 5 \\ 6 \\ 7 \\ 7 \\ 7 \\ 7 \end{bmatrix}.$$

$\lambda_8 = \lambda_9 = 3$ is associated with the <u>eigenvector</u> x_8 and the <u>principal vector</u> x_9 of degree 2 given by

$$x_8 = \frac{1}{8}\begin{bmatrix} 1 \\ 2 \\ 3 \\ 4 \\ 5 \\ 6 \\ 7 \\ 8 \\ 8 \\ 8 \end{bmatrix}, \qquad x_9 = \frac{1}{9}\begin{bmatrix} 1 \\ 2 \\ 3 \\ 4 \\ 5 \\ 6 \\ 7 \\ 8 \\ 9 \\ 9 \end{bmatrix}.$$

Finally, $\lambda_{10} = 1$ is associated with the eigenvector

$$x_{10} = \frac{1}{10}\begin{bmatrix} 1 \\ 2 \\ 3 \\ 4 \\ 5 \\ 6 \\ 7 \\ 8 \\ 9 \\ 10 \end{bmatrix} .$$

These ten normalized vectors (five eigenvectors and five principal vectors) make up the transformation to Jordan canonical form.

Reference: [79, pp. 211-212].

CHAPTER VI

TEST MATRICES: EIGENVALUES AND EIGENVECTORS OF COMPLEX MATRICES

Example 6.1

$$\begin{bmatrix} 1 & -i \\ i & 1 \end{bmatrix}$$

Eigenvalues:

$$\lambda_1 = 2$$
$$\lambda_2 = 0$$

Eigenvectors:

$$x_1 = \begin{bmatrix} 1 \\ i \end{bmatrix}, \qquad x_2 = \begin{bmatrix} -1 \\ i \end{bmatrix}.$$

Reference: [40, p. 142].

Example 6.2

$$\begin{bmatrix} 1 & 1-i \\ 1+i & 1 \end{bmatrix}$$

Eigenvalues:

$$\lambda_1 = 1 + \sqrt{2}$$
$$\lambda_2 = 1 - \sqrt{2}$$

Eigenvectors:

$$x_1 = \begin{bmatrix} \sqrt{2} \\ 1+i \end{bmatrix}, \qquad x_2 = \begin{bmatrix} -\sqrt{2} \\ 1+i \end{bmatrix}.$$

Reference: [40, p. 142].

Example 6.3

$$\begin{bmatrix} 2 & -i & 0 \\ i & 2 & 0 \\ 0 & 0 & 3 \end{bmatrix}$$

Eigenvalues:

$$\lambda_1 = 1$$
$$\lambda_2 = 3$$
$$\lambda_3 = 3$$

Eigenvectors:

$$x_1 = \begin{bmatrix} -1 \\ i \\ 0 \end{bmatrix}, \quad x_2 = \begin{bmatrix} 1 \\ i \\ 0 \end{bmatrix}, \quad x_3 = \begin{bmatrix} 0 \\ 0 \\ 1 \end{bmatrix}.$$

Note: Corresponding to the multiple eigenvalue $\lambda_2 = \lambda_3$, we have a two-dimensional subspace of eigenvectors. x_2 and x_3 are two orthogonal vectors from this subspace.

Reference: [40, p. 99].

Example 6.4

$$\begin{bmatrix} 1+2i & 3+4i & 21+22i \\ 43+44i & 13+14i & 15+16i \\ 5+6i & 7+8i & 25+26i \end{bmatrix}$$

Eigenvalues:

$$\lambda_1 \doteq 6.70088 - 7.87599\ i$$
$$\lambda_2 \doteq 39.7767 + 42.99567\ i$$
$$\lambda_3 \doteq -7.47753 + 6.88032\ i$$

Reference: [15].

Example 6.5

$$\begin{bmatrix} 5+9i & 5+5i & -6-6i & -7-7i \\ 3+3i & 6+10i & -5-5i & -6-6i \\ 2+2i & 3+3i & -1+3i & -5-5i \\ 1+i & 2+2i & -3-3i & 4i \end{bmatrix}$$

Eigenvalues:

$$\lambda_1 = 1 + 5i$$
$$\lambda_2 = 2 + 6i$$
$$\lambda_3 = 3 + 7i$$
$$\lambda_4 = 4 + 8i$$

Right Eigenvectors:

$$x_1 = \begin{bmatrix} 2 \\ 1 \\ 1 \\ 1 \end{bmatrix}, \quad x_2 = \begin{bmatrix} 1 \\ 2 \\ 1 \\ 1 \end{bmatrix}, \quad x_3 = \begin{bmatrix} -1 \\ -1 \\ 0 \\ -1 \end{bmatrix}, \quad x_4 = \begin{bmatrix} -1 \\ -1 \\ -1 \\ 0 \end{bmatrix}.$$

Reference: [60, p. 153].

Example 6.6

$$\begin{bmatrix} 3 & 1 & 0 & 2i \\ 1 & 3 & -2i & 0 \\ 0 & 2i & 1 & 1 \\ -2i & 0 & 1 & 1 \end{bmatrix}$$

Eigenvalues:

$$\lambda_1 = 2 + 2\sqrt{2}$$
$$\lambda_2 = 2 - 2\sqrt{2}$$
$$\lambda_3 = 4$$
$$\lambda_4 = 0$$

Eigenvectors:

$$x_1 = \begin{bmatrix} 1+\sqrt{2} \\ 1+\sqrt{2} \\ i \\ -i \end{bmatrix}, \quad x_2 = \begin{bmatrix} 1 \\ 1 \\ -(1+\sqrt{2})i \\ (1+\sqrt{2})i \end{bmatrix}, \quad x_3 = \begin{bmatrix} -1 \\ 1 \\ i \\ i \end{bmatrix}, \quad x_4 = \begin{bmatrix} 1 \\ -1 \\ i \\ i \end{bmatrix}$$

Reference: Constructed from Examples 6.1 and 6.2 using similarity transformations. See Ortega [44].

Example 6.7

$$\begin{bmatrix} 7 & 3 & 1+2i & -1+2i \\ 3 & 7 & 1-2i & -1-2i \\ 1-2i & 1+2i & 7 & -3 \\ -1-2i & -1+2i & -3 & 7 \end{bmatrix}$$

Eigenvalues:

$$\lambda_1 = 0$$
$$\lambda_2 = 8$$
$$\lambda_3 = 8$$
$$\lambda_4 = 12$$

Eigenvectors:

$$x_1 = \begin{bmatrix} -1 \\ 1 \\ -i \\ -i \end{bmatrix}, \quad x_2 = \begin{bmatrix} -1+i \\ 0 \\ 1 \\ i \end{bmatrix}, \quad x_3 = \begin{bmatrix} i \\ 1 \\ 0 \\ 1+i \end{bmatrix}, \quad x_4 = \begin{bmatrix} 1 \\ 1 \\ 1 \\ -1 \end{bmatrix}.$$

Note: Corresponding to the multiple eigenvalue $\lambda_2 = \lambda_3$, we have a two-dimensional subspace of eigenvectors. x_2 and x_3 are two linearly independent vectors from this subspace.

Reference: Constructed from Example 6.1 using similarity transformations. See Ortega [44].

Example 6.8

Let $A = [a_{ij}]$ be the 5 x 5 Hermitian matrix with the following elements:

i	j	a_{ij} Real Part	a_{ij} Imaginary Part
1	1	-8.45000 00000(-1)	0.00000 00000(0)
1	2	5.20000 00000(0)	1.03000 00000(-1)
1	3	3.01000 00000(-1)	-4.54000 00000(-2)
1	4	-9.60000 00001(0)	9.36000 00000(-1)
1	5	7.33999 99999(-2)	7.26000 00000(0)
2	1	5.20000 00000(0)	-1.03000 00000(-1)
2	2	-6.20000 00000(0)	0.00000 00000(0)
2	3	-3.39000 00000(0)	-4.07000 00000(-1)
2	4	1.22000 00000(-1)	9.10000 00000(-1)
2	5	4.18999 99999(0)	-3.66000 00000(0)
3	1	3.01000 00000(-1)	4.54000 00000(-2)
3	2	-3.39000 00000(0)	4.07000 00000(-1)
3	3	1.90000 00000(-2)	0.00000 00000(0)
3	4	9.35000 00000(-1)	-2.71000 00000(-1)
3	5	-5.72000 00000(-2)	2.82000 00000(0)
4	1	-9.60000 00001(0)	-9.36000 00000(-1)
4	2	1.22000 00000(-1)	-9.10000 00000(-1)
4	3	9.35000 00000(-1)	2.71000 00000(-1)
4	4	7.21000 00000(0)	0.00000 00000(0)
4	5	3.37000 00000(-1)	6.03000 00000(-2)
5	1	7.33999 99999(-2)	-7.26000 00000(0)
5	2	4.18999 99999(0)	3.66000 00000(0)
5	3	-5.72000 00000(-2)	-2.82000 00000(0)
5	4	3.37000 00000(-1)	-6.03000 00000(-2)
5	5	-1.23000 00000(0)	0.00000 00000(0)

Eigenvalues:

$\lambda_1 \doteq 15.18016\ 5225$

$\lambda_2 \doteq 5.67872\ 93543$

$\lambda_3 \doteq -0.83398\ 68001\ 9$

$\lambda_4 \doteq -5.14984\ 56282$

$\lambda_5 \doteq -15.92106\ 2150$

Eigenvectors:

Let $x_j^{(k)}$, j = 1, 2, 3, 4, 5, denote the components of the eigenvector of A corresponding to the eigenvalue λ_k.

k	j	$x_j^{(k)}$ Real Part	$x_j^{(k)}$ Imaginary Part
1	1	6.00205 95562(-1)	0.00000 00000(0)
1	2	1.11536 67096(-1)	-1.01132 04497(-1)
1	3	-8.01643 84119(-3)	4.44447 02492(-2)
1	4	-7.31287 35545(-1)	-9.06611 51813(-2)
1	5	4.60335 97946(-2)	-2.64434 04390(-1)
2	1	-7.60360 24995(-2)	0.00000 00000(0)
2	2	3.20401 41313(-1)	2.35524 37576(-1)
2	3	-5.70992 41394(-1)	-7.46531 68625(-2)
2	4	-2.88357 72893(-1)	1.30852 91636(-1)
2	5	3.00383 89915(-2)	6.35064 97290(-1)
3	1	-3.64692 00756(-1)	0.00000 00000(0)
3	2	-3.67168 77161(-1)	-5.51361 26644(-2)
3	3	3.10174 23865(-1)	5.24760 01509(-1)
3	4	-4.33126 31670(-1)	-1.68050 17325(-1)
3	5	-1.45002 84663(-1)	3.47419 94196(-1)
4	1	4.01976 56816(-1)	0.00000 00000(0)
4	2	3.75061 35332(-1)	4.31499 42367(-1)
4	3	2.81651 56461(-1)	4.57399 51652(-1)
4	4	2.75842 67208(-1)	1.07759 84596(-2)
4	5	-3.54373 44505(-1)	1.45685 63957(-1)
5	1	-5.82568 37541(-1)	0.00000 00000(0)
5	2	5.61740 61319(-1)	1.97319 27448(-1)
5	3	6.64363 95437(-2)	4.26394 75314(-2)
5	4	-2.54317 70782(-1)	2.15142 03800(-3)
5	5	-1.10244 71280(-1)	-4.72290 94206(-1)

Reference: [39].

Example 6.9

$$\begin{bmatrix} 1 + 2i & 3 + 4i & 21 + 22i & 23 + 24i & 41 + 42i \\ 43 + 44i & 13 + 14i & 15 + 16i & 33 + 34i & 35 + 36i \\ 5 + 6i & 7 + 8i & 25 + 26i & 27 + 28i & 45 + 46i \\ 47 + 48i & 17 + 18i & 19 + 20i & 37 + 38i & 39 + 40i \\ 9 + 10i & 11 + 12i & 29 + 30i & 31 + 32i & 49 + 50i \end{bmatrix}$$

Eigenvalues:

$$\lambda_1 \doteq 127.38667\ 077303 + 132.27820\ 320006\ i$$

$$\lambda_2 \doteq 7.07331\ 324882 - 9.55838\ 903704\ i$$

$$\lambda_3 \doteq -9.45998\ 402189 + 7.28018\ 583692\ i$$

$$\lambda_4 \doteq 0.00000\ 000000 + 0.00000\ 000000\ i$$

$$\lambda_5 \doteq 0.00000\ 000000 + 0.00000\ 000000\ i$$

Reference: [15], [26], [48].

Example 6.10

$$
\left[\begin{array}{ccccc}
2 + 3i & 3 + i & 0 & 0 & 0 \\
3 + 2i & -2 - i & 1 + 2i & 0 & 0 \\
5 - 3i & 1 + 2i & 2 + i & -1 + 4i & 0 \\
2 + 6i & -2 + 3i & 3 - i & -4 + 2i & 5 + 5i \\
1 + 4i & 2 + 2i & -3 + 7i & 1 + 5i & 2 - 3i \\
5 - i & 4i & 1 + 5i & -8 - i & 4 + 7i \\
5 + 2i & 1 + 4i & 6 - 5i & 8 + 4i & 4 - 4i \\
-4 - 3i & 7 + 3i & 1 + 6i & 2 - 4i & 3 + i \\
5 & 2 + 2i & 1 + 3i & 1 + i & -4 - 2i \\
5 + 2i & 2 + 6i & 1 - 3i & 7 + 4i & 4 + i
\end{array}\right. \cdots
$$

$$
\cdots \left.\begin{array}{ccccc}
0 & 0 & 0 & 0 & 0 \\
0 & 0 & 0 & 0 & 0 \\
0 & 0 & 0 & 0 & 0 \\
0 & 0 & 0 & 0 & 0 \\
1 + 6i & 0 & 0 & 0 & 0 \\
7 + i & 4 - 2i & 0 & 0 & 0 \\
-1 + 5i & 3 & -4 + 6i & 0 & 0 \\
1 + 2i & 1 + 4i & 6 + 3i & 7 - i & 0 \\
1 + 6i & 1 + 2i & 2 + 5i & i & 3 + 2i \\
-7 & 3 - 3i & 5 - 4i & 6 + 3i & 2 + 5i
\end{array}\right]
$$

Eigenvalues:

$$\lambda_1 \doteq 4.16174868 + 3.13751356i$$

$$\lambda_2 \doteq 5.43644837 - 3.971142582i$$

$$\lambda_3 \doteq 2.38988759 + 7.26807071i$$

$$\lambda_4 \doteq -1.93520144 - 3.975093821i$$

$$\lambda_5 \doteq -2.44755082 + 0.437126175i$$

$$\lambda_6 \doteq -5.27950616 - 2.27596303i$$

$$\lambda_7 \doteq 1.03205812 + 9.29413278i$$

$$\lambda_8 \doteq -4.96687009 - 8.08712475i$$

$$\lambda_9 \doteq 8.81130928 + 1.549382661i$$

$$\lambda_{10} \doteq 10.7976764 + 8.62338151i$$

Reference: [65].

Example 6.11

The matrix displayed on the following pages is a particular case of a class of matrices sometimes referred to as the Dolph-Lewis matrices. These matrices arise in problems in hydrodynamics and are described in [14]. The example included here is of order 20 and corresponds to $\alpha = 0.9$. The eigenvalues, eigenvectors, and condition numbers computed by Wilkinson [70] are also given.

Notice that the diagonal elements of the matrix are complex numbers but the off-diagonal elements are real. Wilkinson computed the eigenvalues extremely accurately, and the complex eigenvalues are presented here to 31 significant digits.

Wilkinson also computed both the left eigenvectors and the right eigenvectors along with the corresponding scalar products (condition numbers). These are included for completeness.

I	J	A(I,J) REAL PART	IMAGINARY PART
1	1	45271 09153 56874 4659(0)	-93634 91190 01533 8361(-3)
2	1	32551 24436 69140 3389(-1)	
3	1	-27544 95253 61182 1681(-3)	
4	1	-14848 81766 37329 1612(-2)	
5	1	13493 90804 53477 7999(-2)	
6	1	-10738 29305 82389 2355(-2)	
7	1	84631 06132 69452 0056(-3)	
8	1	-67551 93899 15280 0441(-3)	
9	1	54834 26066 34832 9186(-3)	
10	1	-45248 41424 42699 5218(-3)	
11	1	37898 74790 52739 2179(-3)	
12	1	-32164 44638 38310 9123(-3)	
13	1	27616 51485 34346 3719(-3)	
14	1	-23955 22133 15619 1558(-3)	
15	1	20967 75424 43387 2104(-3)	
16	1	-18500 44991 44270 0848(-3)	
17	1	16440 47733 86195 3020(-3)	
18	1	-14703 67969 88686 5467(-3)	
19	1	13226 35489 48685 6386(-3)	
20	1	-11959 63495 77477 6675(-3)	

I	J	A(I,J) REAL PART	IMAGINARY PART
1	2	70419 25340 89088 4399(0)	
2	2	58045 85218 42956 5430(0)	-38992 92292 77186 0957(-2)
3	2	69247 79154 36029 4342(-1)	
4	2	-53577 02553 73790 8602(-2)	
5	2	-70315 58634 48046 1478(-3)	
6	2	15374 45263 23862 3738(-2)	
7	2	-15021 45256 38148 1886(-2)	
8	2	13114 30518 39899 2717(-2)	
9	2	-11159 27821 95657 4917(-2)	
10	2	94721 62346 35561 7046(-3)	
11	2	-80809 34130 70309 9072(-3)	
12	2	69461 23030 50234 9138(-3)	
13	2	-60191 87203 49203 7952(-3)	
14	2	52573 78070 38240 1347(-3)	
15	2	-46262 96140 48637 4497(-3)	
16	2	40990 89637 74796 5753(-3)	
17	2	-36549 83520 39139 7178(-3)	
18	2	32778 90245 95404 0438(-3)	
19	2	-29552 91383 84154 0664(-3)	
20	2	26773 75923 66756 8713(-3)	

I	J	A(I,J) REAL PART	IMAGINARY PART
1	3	-41135 35694 77796 5546(0)	
2	3	39207 51772 82094 9554(0)	
3	3	59257 78225 06427 7649(0)	-88344 40261 12556 4575(-2)
4	3	95518 62068 47429 2755(-1)	
5	3	-12315 63789 77000 7133(-1)	
6	3	21654 53741 32692 8139(-2)	
7	3	91682 92490 35799 8759(-4)	
8	3	-68315 93054 81240 1533(-3)	
9	3	81050 54293 99199 7838(-3)	
10	3	-79252 21143 52323 1149(-3)	
11	3	72993 96747 83963 7101(-3)	
12	3	-65765 82964 04406 4283(-3)	
13	3	58803 08599 44410 6221(-3)	
14	3	-52507 06453 80765 1997(-3)	
15	3	46960 03488 78884 6880(-3)	
16	3	-42124 52804 49511 4833(-3)	
17	3	37923 40176 00538 2091(-3)	
18	3	-34272 24437 47334 1823(-3)	
19	3	31091 88983 21688 1752(-3)	
20	3	-28312 57115 75848 9817(-3)	

I	J	A(I,J) REAL PART	IMAGINARY PART
1	4	33830 50374 68671 7987(0)	
2	4	-17120 28067 55900 3830(0)	
3	4	31461 40240 13280 8685(0)	
4	4	59606 79873 82411 9568(0)	-15743 28192 50971 0789(-1)
5	4	11214 67025 95055 1033(0)	
6	4	-17408 36538 37442 3981(-1)	
7	4	44852 23384 10839 4384(-2)	
8	4	-11761 00442 53524 3928(-2)	
9	4	88503 42601 36308 1485(-4)	
10	4	30575 29102 07957 0293(-3)	
11	4	-44478 00019 98879 0154(-3)	
12	4	48075 54541 91239 5507(-3)	
13	4	-47348 31418 31820 8307(-3)	
14	4	44844 89327 29661 4647(-3)	
15	4	-41706 72800 68295 0765(-3)	
16	4	38455 83887 59614 8998(-3)	
17	4	-35327 62602 84474 1195(-3)	
18	4	32420 51207 05300 9421(-3)	
19	4	-29766 56869 57770 9585(-3)	
20	4	27365 98507 90863 8567(-3)	

I	J	A(I,J) REAL PART	IMAGINARY PART
1	5	-30816 64368 51024 6277(0)	
2	5	11878 89436 25986 5761(0)	
3	5	-12003 89862 06054 6875(0)	
4	5	27916 38560 59312 8204(0)	
5	5	59755 48774 00398 2544(0)	-24625 98029 52408 7906(-1)
6	5	12347 24996 61147 5945(0)	
7	5	-21178 48675 69804 1916(-1)	
8	5	63003 83422 05807 5666(-2)	
9	5	-22106 84964 66614 3060(-2)	
10	5	74032 85781 03799 3729(-3)	
11	5	-13349 08629 36014 3095(-3)	
12	5	-13404 16201 85598 7310(-3)	
13	5	25264 82021 49430 2899(-3)	
14	5	-30107 77436 42963 4690(-3)	
15	5	31502 19163 21711 6147(-3)	
16	5	-31175 73942 29061 9016(-3)	
17	5	30004 92943 10156 2560(-3)	
18	5	-28448 71851 26636 1773(-3)	
19	5	26751 57957 12824 9109(-3)	
20	5	-25043 87266 52646 4373(-3)	
1	7	-28357 98151 79109 5734(0)	
2	7	85823 15314 56112 8616(-1)	
3	7	-57750 03787 13011 7416(-1)	
4	7	58531 78212 41915 2260(-1)	
5	7	-86489 89629 00042 5339(-1)	
6	7	24549 26613 71827 1255(0)	
7	7	59878 40294 83795 1660(0)	-48313 07707 35442 6384(-1)
8	7	13785 44121 98066 7114(0)	
9	7	-26342 18125 60409 3075(-1)	
10	7	89195 28336 26419 3058(-2)	
11	7	-37635 90473 68168 8309(-2)	
12	7	17503 45996 69743 3293(-2)	
13	7	-83262 01568 70689 2431(-3)	
14	7	37223 53831 10819 3845(-3)	
15	7	-12680 17440 45224 0407(-3)	
16	7	-88497 37014 31590 5981(-5)	
17	7	84943 42273 47956 0390(-4)	
18	7	-12725 95400 22350 8477(-3)	
19	7	14979 15945 94717 0258(-3)	
20	7	-16047 73697 18137 1972(-3)	

I	J	A(I,J) REAL PART	IMAGINARY PART
1	6	29265 61385 39314 2700(0)	
2	6	-97132 70701 46799 0875(-1)	
3	6	75770 69755 64360 6186(-1)	
4	6	-98389 06768 70942 1158(-1)	
5	6	25876 62786 24534 6069(0)	
6	6	59832 87990 09323 1201(0)	-35482 57378 86130 8098(-1)
7	6	13166 18248 82030 4871(0)	
8	6	-24064 92480 07684 9461(-1)	
9	6	77451 94132 43979 2156(-2)	
10	6	-30586 28761 20954 7520(-2)	
11	6	12872 89953 32494 3781(-2)	
12	6	-50945 89650 98019 6893(-3)	
13	6	13656 84274 78778 1760(-3)	
14	6	50919 58155 38987 5174(-4)	
15	6	-14647 36069 46528 8892(-3)	
16	6	19383 78645 71833 0517(-3)	
17	6	-21497 79556 93029 7986(-3)	
18	6	22152 94753 11399 4375(-3)	
19	6	-21991 29376 09478 8313(-3)	
20	6	21378 16118 19285 8994(-3)	
1	8	27779 62647 37844 4672(0)	
2	8	-79087 00406 55136 1084(-1)	
3	8	48368 84886 02638 2446(-1)	
4	8	-42560 86843 08826 9234(-1)	
5	8	49392 17446 37191 2956(-1)	
6	8	-78971 76407 27758 4076(-1)	
7	8	23615 88831 99095 7260(0)	
8	8	59907 49001 50299 0723(0)	-63117 49480 66473 0072(-1)
9	8	14269 99028 77211 5707(0)	
10	8	-28183 79900 41822 1951(-1)	
11	8	98921 39234 58009 9583(-2)	
12	8	-43582 83767 94233 9182(-2)	
13	8	21468 01620 25615 5729(-2)	
14	8	-11127 48781 45195 5438(-2)	
15	8	57870 68075 73325 9320(-3)	
16	8	-28396 93661 37171 1642(-3)	
17	8	11387 92204 07397 9519(-3)	
18	8	-12902 73530 68522 6433(-4)	
19	8	-47894 16016 13769 3077(-4)	
20	8	84470 36998 40465 5606(-4)	

I	J	A(I,J) REAL PART	IMAGINARY PART
1	9	-27387 88463 17529 6783(0)	
2	9	74719 03786 06319 4275(-1)	
3	9	-42768 39038 35713 8634(-1)	
4	9	34309 83517 31896 4005(-1)	
5	9	-34732 54758 86464 1190(-1)	
6	9	43759 75765 28787 6129(-1)	
7	9	-73791 81124 26996 2311(-1)	
8	9	22923 47662 15085 9833(0)	
9	9	59927 22436 78569 7937(0)	-79895 82978 18899 1547(-1)
10	9	14659 43083 16707 6111(0)	
11	9	-29703 64619 97777 2236(-1)	
12	9	10710 74686 47882 3423(-1)	
13	9	-48665 74192 41964 8170(-2)	
14	9	24899 19657 81897 3064(-2)	
15	9	-13577 43283 73372 5548(-2)	
16	9	76091 55873 06953 9666(-3)	
17	9	-42376 80986 98928 9522(-3)	
18	9	22381 67999 16676 2456(-3)	
19	9	-10111 02849 58849 6611(-3)	
20	9	24069 70634 18338 0753(-4)	

I	J	A(I,J) REAL PART	IMAGINARY PART
1	10	27110 02491 41454 6967(0)	
2	10	-71712 09901 57127 3804(-1)	
3	10	39119 06387 65692 7109(-1)	
4	10	-29398 69672 99133 5392(-1)	
5	10	27221 43614 66467 3805(-1)	
6	10	-30008 47948 71509 0752(-1)	
7	10	39948 30651 20875 8354(-1)	
8	10	-70005 73631 37602 8061(-1)	
9	10	22389 26030 69543 8385(0)	
10	10	59941 23592 97275 5432(0)	-98648 08339 62559 7000(-1)
11	10	14979 25743 46065 5212(0)	
12	10	-30979 23123 27772 3789(-1)	
13	10	11409 24252 10401 4158(-1)	
14	10	-53060 22241 71161 6516(-2)	
15	10	27897 79828 16286 3851(-2)	
16	10	-15738 01811 90092 1166(-2)	
17	10	92286 39537 46855 2589(-3)	
18	10	-54887 63417 81578 9580(-3)	
19	10	32280 41641 70520 3801(-3)	
20	10	-18097 06864 15093 9509(-3)	

I	J	A(I,J) REAL PART	IMAGINARY PART
1	11	-26905 68752 58684 1583(0)	
2	11	69547 69790 17257 6904(-1)	
3	11	-36591 62996 33681 7741(-1)	
4	11	26199 33569 80383 3961(-1)	
5	11	-22771 74545 45199 8711(-1)	
6	11	23017 60064 43798 5420(-1)	
7	11	-26862 14284 04003 3817(-1)	
8	11	37199 82411 71240 8066(-1)	
9	11	-67117 34272 53961 5631(-1)	
10	11	21964 53195 06525 9933(0)	
11	11	59951 54678 82156 3721(0)	-11937 42556 49745 4643(0)
12	11	15246 59376 59144 4016(0)	
13	11	-32065 05347 04327 5833(-1)	
14	11	12012 28459 83132 7200(-1)	
15	11	-56897 61077 51950 6216(-2)	
16	11	30541 06346 09870 6126(-2)	
17	11	-17657 83454 17510 7181(-2)	
18	11	10677 60313 16258 0132(-2)	
19	11	-66149 10089 41084 1465(-3)	
20	11	41238 83154 53426 9124(-3)	

I	J	A(I,J) REAL PART	IMAGINARY PART
1	12	26750 97920 00055 3131(0)	
2	12	-67934 82135 98132 1335(-1)	
3	12	34760 60321 55454 1588(-1)	
4	12	-23980 50436 75005 4359(-1)	
5	12	19879 91412 17023 1342(-1)	
6	12	-18897 21932 82097 5780(-1)	
7	12	20256 06180 54121 7327(-1)	
8	12	-24621 05453 01437 3779(-1)	
9	12	35124 74987 65587 8067(-1)	
10	12	-64840 99384 39726 8295(-1)	
11	12	21618 72256 54721 2601(0)	
12	12	59959 35797 69134 5215(0)	-14207 43465 42358 3984(0)
13	12	15473 37919 47364 8071(0)	
14	12	-33000 54837 01825 1419(-1)	
15	12	12538 22003 49122 2858(-1)	
16	12	-60277 89611 83503 2701(-2)	
17	12	32888 96143 20360 1241(-2)	
18	12	-19375 27587 87013 5903(-2)	
19	12	11981 82770 16747 7429(-2)	
20	12	-76340 71080 24694 0255(-3)	

I	J	A(I,J) REAL PART	IMAGINARY PART
1	13	-26630 99840 28339 3860(0)	
2	13	66699 09134 50717 9260(-1)	
3	13	-33387 28239 75980 2818(-1)	
4	13	22369 30909 56658 1249(-1)	
5	13	-17876 06044 67242 9562(-1)	
6	13	16228 03439 38618 8984(-1)	
7	13	-16381 81717 13501 2150(-1)	
8	13	18310 63814 46123 1232(-1)	
9	13	-22945 46342 45485 0674(-1)	
10	13	33502 83158 94305 7060(-1)	
11	13	-63000 64176 32102 9663(-1)	
12	13	21331 68708 53304 8630(0)	
13	13	59965 41753 41129 3030(0)	-16674 83597 99385 0708(0)
14	13	15668 19753 49783 8974(0)	
15	13	-33814 93268 53454 1130(-1)	
16	13	13000 97641 06944 2034(-1)	
17	13	-63278 48008 36816 4301(-2)	
18	13	34988 67648 88651 6690(-2)	
19	13	-20920 99748 55184 5551(-2)	
20	13	13162 18142 16487 1097(-2)	

I	J	A(I,J) REAL PART	IMAGINARY PART
1	14	26536 05826 19905 4718(0)	
2	14	-65730 47861 45687 1033(-1)	
3	14	32328 43382 28404 5219(-1)	
4	14	-21157 39462 89718 1511(-1)	
5	14	16420 70431 26225 4715(-1)	
6	14	-14381 96073 28623 5332(-1)	
7	14	13880 78124 72805 3808(-1)	
8	14	-14626 96096 85987 2341(-1)	
9	14	16869 06348 91390 8005(-1)	
10	14	-21646 04025 89291 3342(-1)	
11	14	32200 31084 49280 2620(-1)	
12	14	-61481 89073 42851 1620(-1)	
13	14	21089 61232 00654 9835(0)	
14	14	59970 21421 79012 2986(0)	-19339 62944 89502 9068(0)
15	14	15837 35812 45541 5726(0)	
16	14	-34530 30204 40042 0189(-1)	
17	14	13411 31480 87635 6363(-1)	
18	14	-65960 16290 13195 6339(-2)	
19	14	36877 81398 18459 7492(-2)	
20	14	-22319 74445 74907 4221(-2)	

I	J	A(I,J) REAL PART	IMAGINARY PART
1	15	-26459 63169 63434 2194(0)	
2	15	64956 62406 08692 1692(-1)	
3	15	-31493 44772 10044 8608(-1)	
4	15	20220 03196 17986 6791(-1)	
5	15	-15325 05429 34983 9687(-1)	
6	15	13042 45588 83532 8817(-1)	
7	15	-12155 07183 22396 2784(-1)	
8	15	12257 11859 29700 7322(-1)	
9	15	-13337 08188 03086 8769(-1)	
10	15	15759 31231 49126 7681(-1)	
11	15	-20609 21187 51257 6580(-1)	
12	15	31131 31225 10910 0342(-1)	
13	15	-60207 16065 54090 9767(-1)	
14	15	20882 69032 53793 7164(0)	
15	15	59974 07659 88826 7517(0)	-22201 81487 50066 7572(0)
16	15	15985 61834 54394 3405(0)	
17	15	-35163 69825 22904 8729(-1)	
18	15	13777 67883 24117 6605(-1)	
19	15	-68371 39895 65148 9496(-2)	
20	15	38586 73240 75661 5996(-2)	

I	J	A(I,J) REAL PART	IMAGINARY PART
1	16	26397 19359 57670 2118(0)	
2	16	-64328 24861 25826 8356(-1)	
3	16	30822 52689 63724 3748(-1)	
4	16	-19478 40675 71163 1775(-1)	
5	16	14476 48776 22947 0968(-1)	
6	16	-12034 27882 86507 1297(-1)	
7	16	10904 80294 07858 8486(-1)	
8	16	-10626 04016 62439 1079(-1)	
9	16	11072 28884 93910 4319(-1)	
10	16	-12350 82198 40005 0402(-1)	
11	16	14879 22994 41993 2365(-1)	
12	16	-19762 82452 23313 5700(-1)	
13	16	30238 20067 75587 7972(-1)	
14	16	-59121 96729 33220 8633(-1)	
15	16	20703 78065 10925 2930(0)	
16	16	59977 23266 48235 3210(0)	-25261 39244 43721 7712(0)
17	16	16116 63084 47718 6203(0)	
18	16	-35728 45179 58760 2615(-1)	
19	16	14106 78809 04302 0010(-1)	
20	16	-70551 28284 26435 5898(-2)	

I	J	A(I,J) REAL PART	IMAGINARY PART
1	17	-26345 52158 41531 7535(0)	
2	17	63810 83093 58358 3832(-1)	
3	17	-30274 79653 24670 0764(-1)	
4	17	18880 51629 06646 7285(-1)	
5	17	-13804 02047 65275 1207(-1)	
6	17	11253 26985 49330 2345(-1)	
7	17	-99646 24419 80838 7756(-2)	
8	17	94464 52138 94546 0320(-2)	
9	17	-95174 32772 55445 7188(-2)	
10	17	10171 94066 66234 1356(-1)	
11	17	-11573 14667 94386 5061(-1)	
12	17	14164 49493 26291 6803(-1)	
13	17	-19058 88994 22436 9526(-1)	
14	17	29480 85777 46152 8778(-1)	
15	17	-58186 94597 11015 2245(-1)	
16	17	20547 55650 46072 0062(0)	
17	17	59979 84409 33227 5391(0)	-28518 36159 82532 5012(0)
18	17	16233 23280 36427 4979(0)	
19	17	-36235 13923 95794 3916(-1)	
20	17	14404 05915 49128 2940(-1)	

I	J	A(I,J) REAL PART	IMAGINARY PART
1	19	-26265 70798 45666 8854(0)	
2	19	63016 21533 93030 1666(-1)	
3	19	-29441 87121 46580 2193(-1)	
4	19	17984 20283 01209 2113(-1)	
5	19	-12815 17464 66755 8670(-1)	
6	19	10133 35841 70788 5265(-1)	
7	19	-86593 25889 31173 0862(-2)	
8	19	78750 42036 17572 7844(-2)	
9	19	-75533 80390 16351 1038(-2)	
10	19	76046 61397 63593 6737(-2)	
11	19	-80258 62742 21539 4974(-2)	
12	19	88976 83001 12336 8740(-2)	
13	19	-10426 42700 48588 5143(-1)	
14	19	13074 61038 23184 9670(-1)	
15	19	-17955 34067 78723 0015(-1)	
16	19	28265 94957 15051 8894(-1)	
17	19	-56657 80929 84855 1750(-1)	
18	19	20287 81957 92436 5997(0)	
19	19	59983 87932 77740 4785(0)	-35624 47726 72653 1982(0)
20	19	16431 78015 94734 1919(0)	

I	J	A(I,J) REAL PART	IMAGINARY PART
1	18	26302 27282 64331 8176(0)	
2	18	-63379 55407 79829 0253(-1)	
3	18	29821 49692 25227 8328(-1)	
4	18	-18390 79870 84239 7213(-1)	
5	18	13260 89620 12454 8674(-1)	
6	18	-10633 96979 58901 5245(-1)	
7	18	92365 90936 77997 5891(-2)	
8	18	-85605 20596 80223 4650(-2)	
9	18	83951 09674 89331 9607(-2)	
10	18	-86797 12424 05474 1859(-2)	
11	18	94657 69398 95749 0921(-2)	
12	18	-10944 66028 73504 1618(-1)	
13	18	13572 65945 98963 8567(-1)	
14	18	-18464 26725 38757 3242(-1)	
15	18	28830 50311 35678 2913(-1)	
16	18	-57372 92440 60814 3806(-1)	
17	18	20409 95229 03561 5921(0)	
18	18	59982 03083 87279 5105(0)	-31972 72308 17079 5441(0)
19	18	16337 67466 99213 9816(0)	
20	18	-36692 29894 87648 0103(-1)	

I	J	A(I,J) REAL PART	IMAGINARY PART
1	20	26234 51724 64847 5647(0)	
2	20	-62707 18947 05295 5627(-1)	
3	20	29120 62196 99233 7704(-1)	
4	20	-17642 62723 73646 4977(-1)	
5	20	12444 36891 28383 9941(-1)	
6	20	-97220 99523 06002 3785(-2)	
7	20	81925 87876 69241 4284(-2)	
8	20	-73317 66129 93702 2924(-2)	
9	20	69027 69091 53908 4911(-2)	
10	20	-67995 90366 42685 5326(-2)	
11	20	69904 54625 33831 5964(-2)	
12	20	-75021 08055 63256 1445(-2)	
13	20	84311 18563 74889 6122(-2)	
14	20	-99919 05535 57127 7142(-2)	
15	20	12649 73520 30128 2406(-1)	
16	20	-17514 83767 29339 3612(-1)	
17	20	27771 26268 48369 8368(-1)	
18	20	-56024 61285 88914 8712(-1)	
19	20	20178 70470 88146 2097(0)	
20	20	59985 45661 56864 1663(0)	-39473 62303 73382 5684(0)

EIGENVALUES

	REAL PART	IMAGINARY PART
1	0.190038863 3975896246 8207230665 1	-0.215008457 7602084027 4025322840 0
2	0.183452778 3718224009 5583109292 2	-0.176568538 5870739008 8862618827 6
3	0.592930305 6182862874 7122654154 7	-0.140888947 3533186860 0691745139 4
4	0.535220447 6329715975 0306764219 8	-0.141263130 4984021724 9845212029 6
5	0.481285813 7770657576 1517590063 5	-0.141103339 8793303898 5118564508 6
6	0.413414690 1693897078 8280776578 0	-0.141327999 0715091091 6275814065 7
7	0.777764085 9441888714 8733245007 7	-0.186925814 3349321476 9116030292 1
8	0.347760817 9682195079 9977073364 7	-0.141957761 5173601794 6905493933 6
9	0.763932870 7907607449 8108880248 1	-0.138787096 5640070089 9986346194 8
10	0.785876563 7149536517 5074463359 6	-0.113440005 3966406666 4357187600 8
11	0.305509379 1347431406 2009523722 4	-0.124485416 3169993538 0872511197 1
12	0.803814688 3346488051 3144306437 6	-0.247603569 0718583082 6416959528 7
13	0.733969743 5409688619 5522620752 1	-0.140217031 5179807143 2318078465 7
14	0.812778463 0935692249 1572885856 4	-0.086714318 4069569104 7463400092 8
15	0.690661404 9467707358 0128713971 7	-0.140154726 6558511635 8514821558 5
16	0.203803654 6240054096 0888764864 0	0.003326547 8966812518 1715340224 6
17	0.841204035 5059620626 9679400580 7	-0.323979871 2081584755 7403882355 7
18	0.839620250 9906857460 1997623006 9	-0.060060466 7273840046 8752892080 2
19	0.642036654 7350083652 5943022874 0	-0.141049064 8312920122 7861217658 9
20	0.866457510 3803274381 6689632229 9	-0.033404574 4580333028 8428515962 7

CONDITION NUMBERS

	REAL PART	IMAG. PART
1	-0.032875366	-0.088548428
2	-0.013546439	-0.034630907
3	-0.002773871	-0.002936017
4	0.004506368	-0.000781866
5	-0.002554063	0.004554858
6	-0.003290467	-0.004946580
7	0.018535904	-0.014170278
8	0.005262802	-0.000995753
9	0.003007399	-0.000360438
10	-0.005432201	-0.001265974
11	-0.004813871	0.004723745
12	0.050142980	-0.220685641
13	-0.001121502	0.001786629
14	0.017361091	0.009402279
15	-0.001256142	-0.002492914
16	0.044986778	0.054156178
17	1.226697223	0.039876322
18	-0.079497763	-0.002272420
19	-0.000431942	0.003393100
20	0.451106514	-0.059596278

RIGHT EIGENVECTORS

X(1) REAL PART	X(1) IMAG. PART	X(2) REAL PART	X(2) IMAG. PART	X(3) REAL PART	X(3) IMAG. PART	X(4) REAL PART	X(4) IMAG. PART
1000000001	1	1000000001	1	1000000001	1	1000000001	1
-343173058	-46556047	-292689402	-45788585	126511539	-246683425	33025476	-213403695
223800116	72090688	162998889	61712495	-192776038	-1017662	-156829589	59657347
-172038116	-98591565	-108303507	-73716405	22803844	141512905	84455657	80468137
133408556	126380474	72607294	83183362	128749561	-59588809	44078954	-98845360
-92732050	-152459542	-42264503	-88975747	-89865480	-100206702	-106084482	5461155
44620161	171794890	13504577	89342304	-72588994	111364345	48143848	88458395
11101530	-178187501	13608831	-82742131	124465230	39984352	58691199	-77519408
-70057618	165848755	-37028045	68591094	8201110	-127070273	-90181580	-20660902
123914567	-131683205	53885148	-47878693	-119615308	22805656	15499578	84998010
-162230220	77606638	-61689595	23380148	47116329	100239598	65704343	-44088586
175806598	-11693326	59488571	790623	74406250	-60431923	-58284065	-36429477
-160645009	-53043137	-48583862	-20299445	-62259442	-47191231	-8100577	55527338
120651391	102523061	32396267	32016256	-23992561	54927732	41722075	-11348574
-67238039	-126827362	-15404925	-35098194	42434147	6697609	-19453854	-24280573
15360455	124187608	1597197	31150545	-3087789	-27813223	-9917543	18702776
22287565	-101354549	6806007	-23351640	-15671643	6344330	13592343	432770
-39782305	69943669	-9719646	15019760	5856764	6983054	-3399492	-6701263
38701373	-40837257	8576025	-8383528	2825868	-3638149	-2498111	2848965
-24898511	19579046	-5173751	4037921	-2740017	-362963	2200369	275128

X(5) REAL PART	X(5) IMAG. PART	X(6) REAL PART	X(6) IMAG. PART	X(7) REAL PART	X(7) IMAG. PART	X(8) REAL PART	X(8) IMAG. PART
1000000001	1	1000000001	1	-43009381	176830391	1000000001	1
-41556560	-184016966	-120418415	-149126998	67671560	107540437	-180688672	-118303478
-109645937	93412946	-42343798	111547786	107368455	-24687863	21101418	110685554
102776480	20797527	85028740	-39533372	4511588	-130207277	41921722	-74694925
-24254367	-82708406	-67891980	-27956986	-161064977	-49452456	-63395715	27684171
-57757229	62641014	17047881	62955218	-122290884	192234867	55070826	16526642
81936649	11054818	36467342	-53196507	206032672	236574799	-25919358	-44821403
-33911015	-69876373	-62138621	10213789	394169069	-170468251	-10489949	49027513
-38258605	61910179	46992969	35211435	-42502754	-571486021	38251578	-30489390
66809914	-457075	-5313194	-52747530	-702605567	-209343315	-45955025	618131
-31299908	-50766260	-31495805	34360341	-559625744	684804638	32818340	24598122
-23821468	46357074	39584112	956888	436474359	885747957	-8938499	-33590114
43443375	-3868709	-20417678	-25250324	1000000001	1	-11504306	25558752
-20852255	-26735575	-4875549	25480614	414343464	-797444538	19388384	-9310747
-8292353	23064035	17153413	-10312953	-402483460	-574760983	-15163167	-4247260
16283312	-3589148	-13843906	-3572241	-466216829	75483172	5885963	9489516
-7403669	-7387615	4649038	7706159	-66271812	271977273	1446557	-7758948
-1551366	5952936	1472047	-4789947	119618027	82432379	-4191919	3684337
3693423	-1231153	-2723722	1004974	47639652	-43981285	3592877	-781285
-2101197	-643889	1634790	513372	-11421594	-29905391	-1880376	-120981

X(9)

REAL PART	IMAG. PART
1000000001	1
491056475	-353984461
-82728487	-395658056
-400952936	-21293272
-151156761	400856065
376518427	289539833
443323393	-304130069
-161259989	-576455320
-643326075	-48353157
-284125845	596614330
425843776	466666297
521901396	-184576065
27566821	-436715624
-277982064	-134906589
-140872615	133835711
46267676	98563968
53904914	-9286440
1178552	-24813308
-9521422	-1677871
-1578477	3959107

X(10)

REAL PART	IMAG. PART
1000000001	1
565067642	-301987882
61281460	-404781081
-318810693	-177484023
-301317890	220088350
75709722	376021917
392052869	94346943
265743064	-318238788
-157465796	-375674690
-375220380	-37412672
-183655546	262526826
106233380	220872222
167480171	7476260
49163113	-89674386
-34059607	-44179643
-26230794	8026127
42726	12160971
4709133	1126955
699526	-1575627
-596662	-368523

X(11)

REAL PART	IMAG. PART
1000000001	1
-204667163	-87608526
54467310	88675554
5740727	-70868733
-33518732	44065702
41192912	-14812008
-33789319	-10299572
16736467	25652603
2905045	-28581218
-18031171	20680842
24117148	-7273460
-20985319	-5079149
12361011	11721101
-3375473	-11772142
-2316384	7624077
3932670	-2716574
-2913891	-617531
1236823	1882412
-137609	-1768631
-134478	1100217

X(12)

REAL PART	IMAG. PART
-201170	4700151
-5639146	-2930513
-1947934	6305190
6651140	4425413
10612911	-8610318
-10922565	-21379675
-40380471	9222581
-4637022	72286210
116756926	45899630
137987422	-157909208
-150535079	-299374143
-504979572	17476485
-303907633	636299676
496886550	741623736
1000000001	1
569579784	-766557027
-197774297	-713392480
-428795525	-139961396
-167150640	195330656
21713177	138482458

X(13)

REAL PART	IMAG. PART
1000000001	1
416294493	-335920746
-140245064	-299102691
-315938306	86619845
24715583	330661843
355885203	36886305
117436987	-372901830
-369258980	-207484751
-299084156	330226826
250815197	370436522
399014898	-142009846
-29729333	-370157008
-292828184	-54503298
-92534828	195725563
110437560	88840143
64429166	-52767829
-21723105	-38021694
-19505772	8066219
2944133	8118306
3868642	-1021753

X(14)

REAL PART	IMAG. PART
1000000001	1
653328313	-242601304
242754045	-388589089
-150200664	-315492420
-332768972	-36012933
-216708610	242790299
70118380	300243272
264621959	110557460
219409858	-124639560
31118582	-199036420
-98494420	-104517105
-90852291	11829552
-22634164	47214948
15196253	22205283
12474454	-1720276
1319923	-5168604
-1683785	-1115384
-566489	461760
146187	214246
40662	-29220

X(15)

REAL PART	IMAG. PART
1000000001	1
317701320	-306874630
-185654794	-180194053
-188762908	160500371
151717155	182422929
195515922	-150538643
-147850447	-209627109
-227730405	142980301
131447817	239893363
243560598	-111557130
-84505499	-233913698
-210029614	55532687
29748697	172965160
129724447	-11252218
-1092865	-87996042
-54289159	-2024068
-1744804	29996776
15251043	543174
47740	-6481000
-3472098	290437

X(16)

REAL PART	IMAG. PART
1000000001	1
-236370670	-2048460
101672929	1762041
-52777861	-966224
29625236	-23317
-17220960	1018970
10219619	-1872348
-6228004	2479863
3981377	-2795730
-2734576	2833913
2029852	-2655546
-1593974	2345520
1279081	-1987147
-1018329	1643314
789658	-1348995
-589070	1114000
414562	-931072
-259702	782876
114198	-642390
24310	456410

X(17) REAL PART	X(17) IMAG. PART	X(18) REAL PART	X(18) IMAG. PART	X(19) REAL PART	X(19) IMAG. PART	X(20) REAL PART	X(20) IMAG. PART
21996163	-21236584	1000000001	1	1000000001	1	1000000001	1
-4879693	5078259	742209469	-175552447	216639629	-276791892	831645305	-100395211
2198166	-2373841	442599672	-325416834	-203582194	-77540525	656515253	-207564440
-1380930	1340364	109692669	-367260222	-67459193	173913254	446350016	-284172260
746465	-782155	-155881625	-259803018	176839874	35264414	235706375	-296480725
-301306	1109105	-267081649	-61239303	14559849	-176705525	66044998	-243409338
1507385	-1260689	-211529079	114367467	-180095731	7807444	-33120910	-152436661
-1378103	-1988495	-70991673	176703264	26764742	179886886	-62321656	-64658078
-5799663	53319	44017764	127382076	176444838	-45528968	-47188150	-8444397
-3216193	14275846	75774728	39474585	-60394625	-164436019	-20674062	11860859
26782061	15338451	46371874	-16000592	-145624359	68938653	-2969964	10598398
51311525	-38430280	9000729	-23923692	70208612	121079369	2531773	3951539
-26085277	-126652199	-6634700	-10123447	94037282	-65348034	1887656	160721
-229936600	-69075110	-5303290	90769	-55724126	-66408167	410960	-560597
-313690583	265261386	-929945	1896118	-42327839	43010651	-83598	-211165
29733973	638497109	467247	585285	30194455	23746785	-87180	-8135
664184699	561823635	264290	-74546	11721884	-19177255	1327	26495
1000000001	1	-20682	-91257	-11374973	-4298142	-3622	-915
756381823	-485447277	-5549	6794	-1048161	5324076	10208	3493
282957402	-394246931	-21851	4685	2954113	-119299	-10268	-967

LEFT EIGENVECTORS

Y(1) REAL PART	Y(1) IMAG. PART	Y(2) REAL PART	Y(2) IMAG. PART	Y(3) REAL PART	Y(3) IMAG. PART	Y(4) REAL PART	Y(4) IMAG. PART
779344	2955195	-1080976	-4414982	-165132	-344102	419936	85613
-5226967	-8389084	7275200	13525423	-1744148	-166950	978118	-1737659
12897237	10827602	-17687182	-19588389	2462051	6347711	-8128811	684711
-18899304	-10247838	24075306	20411758	13415807	-8843249	9372455	15540284
18435362	8744397	-17556662	-15224916	-28663045	-25093806	24987650	-32247762
-6900855	-7352593	-8772972	421571	-31293784	67940379	-79500601	-15614959
-19948623	3065628	56546752	33337644	135720716	24323919	32741121	138542817
63922885	13318197	-117398428	-100380075	-21839688	-232334059	189146694	-144278987
-119748168	-56908815	168616908	213701664	-341677687	123530713	-322509509	-171821999
169411604	143629149	-173669892	-373271984	296994782	435237382	-39222984	518056116
-180024835	-279177186	92533734	553625551	465372766	-524813371	652198855	-228812420
111260875	445305423	97538535	-700067012	-763878737	-402315921	-577730006	-618100711
64887444	-592673563	-379079795	742767279	-232001637	944332516	-383625685	876070003
-338735356	651665752	685475713	-629074017	1000000001	1	1000000001	1
649131872	-564609405	-919976073	359112843	-234497403	-910034836	-386740907	-867770171
-898909629	325544935	1000000001	1	-691724467	389232588	-556814204	614824702
1000000001	1	-902470653	-337363749	436105306	436935841	640397179	188339636
-922504126	-295740005	678675123	548753519	186094728	-379575786	-83240951	-454615536
709948462	446552046	-425055538	-580057195	-248943186	-55743835	-264012993	161764073
-436757439	-380819310	223098270	427082906	45120978	178485521	212838514	74523157

Y(5)

REAL PART	IMAG. PART
-395772	545016
1857136	785661
288657	-8680246
-17018473	13591286
40747412	14608473
-16842886	-83544480
-106283716	107669259
237531178	50493893
-125364982	-347396246
-320887771	414963648
683430454	66763567
-367911926	-754741694
-508781240	810519020
1000000001	1
-506408178	-819674575
-377206628	779076923
691893491	-103511191
-326167450	-374417566
-117321732	343744158
223750421	-131151468

Y(6)

REAL PART	IMAG. PART
-237140	-659981
-1693387	1579531
8885463	2981004
-12557157	-21115101
-11950643	46401716
79344789	-37918142
-143608968	-60481165
87131445	235916412
171831374	-329938980
-504153029	123835548
562736542	388778697
-105938867	-829750892
-622608515	712219153
1000000001	1
-658790110	-737954604
-82675873	890997373
576181750	-439519443
-536129697	-120600416
216087359	372508176
1520403	-295687639

Y(7)

REAL PART	IMAG. PART
-12587	-4136
-40301	-10126
-96014	237583
749088	560694
2310469	-1807880
-3163970	-7658328
-21039680	2236089
-9938690	48411610
90839105	54056422
161593995	-128116327
-98855481	-350614364
-569450105	-87846433
-461092987	658490660
449385549	861626037
1000000001	1
379341741	-784565993
-434537773	-494497475
-402390470	149842790
31915010	235834314
139879538	35076947

Y(8)

REAL PART	IMAG. PART
829420	117031
-1742028	-2215198
-3229079	8209617
21618689	-12161202
-51910440	-4553354
68260084	62476578
-19517925	-152827841
-133963975	206988823
352308006	-114136618
-480560784	-184227484
327748165	579330924
149147712	-796556643
-719963985	590830573
1000000001	1
-769374680	-619276507
185428616	870763254
355262479	-654235466
-566741545	212793880
466759315	124025375
-254227162	-193482001

Y(9)

REAL PART	IMAG. PART
9351	65996
462689	581563
2491693	-692417
-101876	-8139261
-20328325	-6267280
-27452408	40729937
63560494	81088579
184825553	-64458699
4201521	-338305223
-490416025	-196645592
-513833523	531840288
359766396	844799592
1000000001	1
359941550	-878611888
-576716956	-535692927
-500893847	274938415
84831404	355404206
208375889	-2896373
11610185	-100745766
-51941127	-16250522

Y(10)

REAL PART	IMAG. PART
69638	-149340
-71577	-1933132
-5320531	-3612033
-14836772	10494114
8212606	44868578
94516835	24072620
120025609	-147192125
-137863070	-303667812
-524722784	-26999223
-385807247	632798315
451576058	804232544
1000000001	1
445363129	-829644150
-449190267	-628846368
-544251629	115152984
-52737078	349442818
179763839	86353508
63950087	-77908179
-29694807	-33123242
-17851830	10723934

Y(11)

REAL PART	IMAG. PART
-211275	1196173
3802563	-4711586
-15799046	5142686
34594067	10904948
-42697099	-56910780
5614365	127662747
108218732	-179882829
-287085668	136152799
443341157	67836194
-434655108	-407872840
158803749	730920023
332238725	-818397165
-809752370	543295600
1000000001	1
-787288178	-530518944
307810778	778501468
160577093	-676950176
-408357207	377613966
410034018	-100925964
-264851997	-20812035

Y(12)

REAL PART	IMAG. PART
15855	-13281
-14527	31155
-11411	-20080
-5462	44140
143519	13283
213916	-421961
-1199742	-1047420
-3977178	2590756
3438483	12658373
33363980	2663198
35188886	-70001431
-104385185	-130758426
-315680244	61830367
-183444599	516216338
482390656	649188229
1000000001	1
606776823	-779709574
-240555987	-717389137
-534148901	-36797712
-251111264	255762993

Y(13)

REAL PART	IMAG. PART
-90510	15222
-634609	737414
1810197	2524429
8754649	-4295456
-6152711	-23374616
-54170721	4337334
-13187320	109332028
192602942	65396129
179990520	-292103926
-369429466	-373845343
-629639863	366032772
236155747	873047837
1000000001	1
248409957	-945553451
-742393852	-398961716
-414027250	490273303
278670230	328219733
217719861	-142411464
-66045680	-113581541
-69297361	32441218

Y(14)

REAL PART	IMAG. PART
-268629	289819
-2000575	4685844
4443479	16784019
40045102	12900046
75305943	-64257937
-17402895	-199118682
-303432974	-176380093
-509664095	227027244
-185886034	764085439
586460252	756233499
1000000001	1
528716177	-731827254
-255071048	-663497443
-486545768	-76110643
-174809272	246903547
87861939	140626353
79405798	-18559712
1176997	-36156266
-13960513	-2985195
-2294576	5325035

Y(15)

REAL PART	IMAG. PART
21462	-173668
-726948	-1115652
-3377950	2453935
6459984	10756223
25374309	-14697029
-29540942	-56313247
-110997489	52884034
83932812	202812087
338068667	-116287217
-136778936	-516829150
-715918147	128269382
80288944	895206076
1000000001	1
83922916	-996031097
-879582318	-138505365
-145267528	694804975
487254954	114303305
66633620	-314599686
-167465028	-27483546
-9546336	108187687

Y(16)

REAL PART	IMAG. PART
17108793	-7370070
-74006528	1143965
149747119	67970522
-191544740	-216251238
140118905	417138199
38633699	-602526971
-327212210	686790288
651645045	-606381104
-906303254	356060332
1000000001	1
-899765563	-352530108
646691692	596173825
-333790070	-675886229
59570471	604484725
114396675	-444322332
-179267307	268955189
165579513	-130006014
-116250801	45242904
64633791	-6654897
-26484848	-3576415

Y(17)

REAL PART	IMAG. PART
-30786	-81
51054	-20138
-26640	29429
21460	-14216
-27050	13648
21714	-10068
11301	12849
24796	-101565
-301269	-155392
-863342	859235
1756961	3435560
11333671	-835140
11302621	-29396024
-51195527	-59793005
-175999210	23265066
-178916397	309579472
182172118	592917473
754246977	484055881
1000000001	1
616215537	-244482173

Y(18)

REAL PART	IMAG. PART
302848	-893314
1354431	-13213695
-14770693	-47284865
-89341883	-64263045
-208871635	44090655
-211945665	339953546
105658334	643888305
666781728	577827560
1000000001	1
702960719	-665202173
-588809	-812822782
-461146434	-383641381
-380669380	96582210
-70409761	218717868
84055005	86794545
53879301	-19685105
-71860	-24679119
-9297508	-2608823
-1538393	3042624
955720	746504

Y(19)

REAL PART	IMAG. PART
-144162	175598
309617	1625362
4609943	-2542994
-7038523	-12508527
-25986801	21103448
48603696	48500274
79540889	-103140008
-191483355	-116934730
-153766208	324320516
494496791	178569034
180510520	-686755341
-861771523	-149203646
-86771902	979757893
1000000001	1
-85667533	-916185018
-744050276	156693767
183132711	534540466
331163131	-186757904
-129453067	-168960125
-89026133	117646296

Y(20)

REAL PART	IMAG. PART
-468350	2182138
-3262871	34292868
13547380	145992192
123928734	337007575
398606025	487642185
772956892	401850984
1000000001	1
835682126	-515610114
321127477	-777419160
-193097054	-606988643
-379918593	-202451444
-245002865	91407736
-41521390	137385401
46053768	54656348
31775460	-6331461
3098705	-13039334
-4214174	-2822826
-1349699	1131532
265042	488306
166417	-49920

CHAPTER VII

TEST MATRICES: EIGENVALUES AND EIGENVECTORS OF TRIDIAGONAL MATRICES

Example 7.1

$$\text{Let } A = \begin{bmatrix} \alpha_1 & \beta_2 & & & \\ 1 & \alpha_2 & \beta_3 & & \\ & \cdots & \cdots & \cdots & \\ & & 1 & \alpha_{15} & \beta_{16} \\ & & & 1 & \alpha_{16} \end{bmatrix}$$

where the α_i and β_i are given by

α_i	β_i
3 + 2i	.....
-1 - i	4 - i
3 - 4i	2 + 4i
2 + 3i	3 + i
-5 + i	3 - 2i
1 + 2i	2 - 2i
5 + 2i	2 + 3i
-2 + i	1 + 3i
1 - 2i	-2 + 2i
-1 - 4i	3 + 3i
2 + i	-1 + 5i
1 - 5i	4 + 3i
3 + i	1 - 6i
2 + 4i	2 + i
-4 + 3i	5 - i
1 - 5i	-3 - 4i

Eigenvalues:

2.06853152 - 2.054443045i	1.57598142 - 3.83032770i
2.40341933 + 2.081055512i	3.28048252 + 3.27566163i
2.72491267 - 2.378378451i	1.19252750 - 5.44399752i
2.45640400 + 0.6319368611i	3.55339888 + 1.26465631i
2.27740066 + 1.448268501i	-2.45560768 - 4.69290496i
0.812811959 + 1.335511351i	-4.89673115 + 3.62210856i
-1.38565721 - 1.387560511i	-5.65716067 + 1.63200082i
-2.72480368 + 0.6570645461i	5.77408994 + 2.83933591i

Reference: [65].

<u>Example 7.2</u>

$$\begin{bmatrix} 10 & 1 \\ 1 & 9 & 1 \\ & 1 & 8 & 1 \\ && \ddots & \ddots & \ddots \\ &&& 1 & 1 & 1 \\ &&&& 1 & 0 & 1 \\ &&&&& 1 & 1 & 1 \\ &&&&&& \ddots & \ddots & \ddots \\ &&&&&&& 1 & 8 & 1 \\ &&&&&&&& 1 & 9 & 1 \\ &&&&&&&&& 1 & 10 \end{bmatrix}$$

Eigenvalues:

10.74619 42	4.99978 25
10.74619 42	4.00435 40
9.21067 86	3.99604 82
9.21067 86	3.04309 93
8.03894 11	2.96105 89
8.03894 11	2.13020 92
7.00395 22	1.78932 14
7.00395 18	0.94753 44
6.00023 40	0.25380 58
6.00021 75	-1.12544 15
5.00024 44	

Reference: [62, p. 309].

Example 7.3

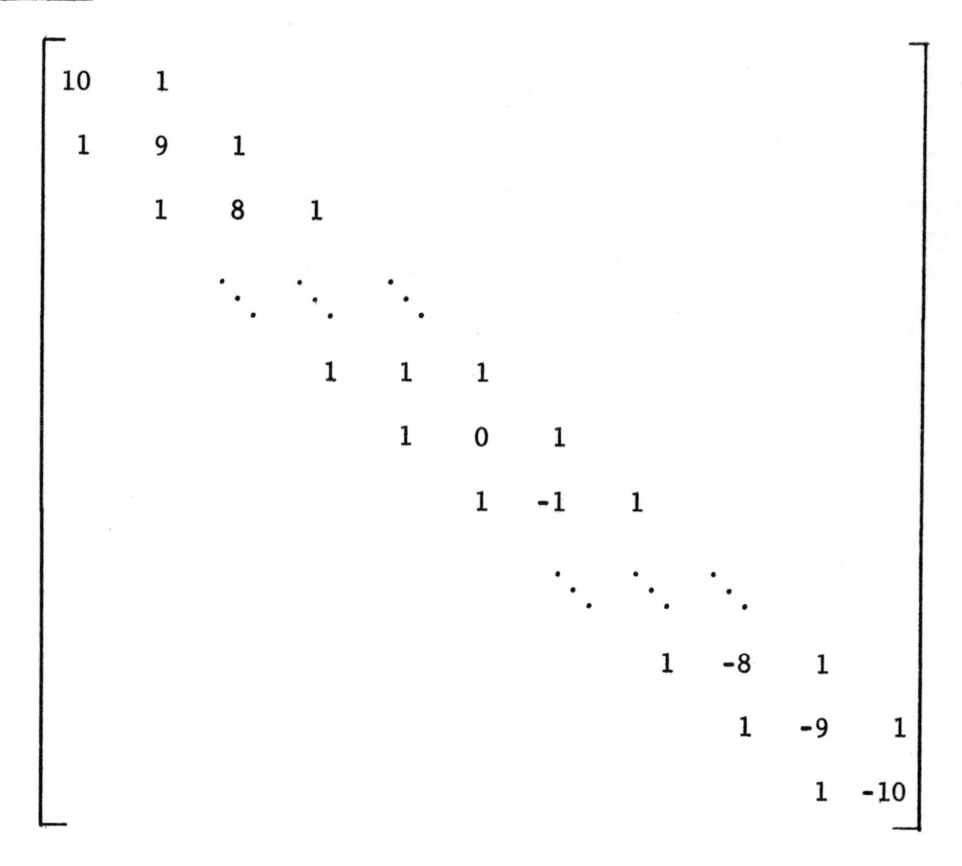

Eigenvalues:

±10.74619 42
± 9.21067 86
± 8.03894 11
± 7.00395 20
± 6.00022 57
± 5.00000 82
± 4.00000 02
± 3.00000 00
± 2.00000 00
± 1.00000 00
0.00000 00

Reference: [62, p. 309].

Example 7.4

$$\begin{bmatrix} a & b & & & & \\ b & a & b & & & \\ & b & a & b & & \\ & \cdots & \cdots & \cdots & \cdots & \\ & & & b & a & b \\ & & & & b & a \end{bmatrix}, \quad n \times n.$$

Eigenvalues:

$$\lambda_k = a + 2b \cos \frac{k\pi}{n+1}, \quad k = 1, 2, \ldots, n.$$

Eigenvectors:

$$x^{(k)} = \begin{bmatrix} x_1^{(k)} \\ x_2^{(k)} \\ \vdots \\ x_n^{(k)} \end{bmatrix}, \text{ where } x_j^{(k)} = \left(\frac{2}{n+1}\right)^{\frac{1}{2}} \sin \frac{kj\pi}{n+1},$$

$$j = 1, 2, \ldots, n; \; k = 1, 2, \ldots, n.$$

Reference: [18, pp. 20, 25].

Example 7.5

$$\begin{bmatrix} (a-b) & b & & & & \\ b & a & b & & & \\ & b & a & b & & \\ & \cdots & \cdots & \cdots & \cdots & \\ & & & b & a & b \\ & & & & b & a \end{bmatrix}, \quad n \times n.$$

Eigenvalues:

$$\lambda_k = a + 2b \cos \frac{2k\pi}{2n+1}, \quad k = 1, 2, \ldots, n.$$

Reference: [18, pp. 27-28].

Example 7.6

$$\begin{bmatrix} (a-b) & b & & & & \\ b & a & b & & & \\ & b & a & b & & \\ \cdots & \cdots & \cdots & \cdots & \cdots & \cdots \\ & & & b & a & b \\ & & & & b & (a+b) \end{bmatrix}, \quad n \times n.$$

Eigenvalues:

$$\lambda_k = a + 2b \cos \frac{(2k-1)\pi}{2n}, \quad k = 1, 2, \ldots, n.$$

Reference: [18, pp. 27-28].

Example 7.7

$$\begin{bmatrix} (a+b) & b & & & & \\ b & a & b & & & \\ & b & a & b & & \\ \cdots & \cdots & \cdots & \cdots & \cdots & \cdots \\ & & & b & a & b \\ & & & & b & (a+b) \end{bmatrix}, \quad n \times n.$$

Eigenvalues:

$$\lambda_k = a + 2b \cos \frac{k\pi}{n}, \quad k = 1, 2, \ldots, n.$$

Reference: [18, p. 29].

Example 7.8

$$\begin{bmatrix} a & 1 & & & & \\ 1 & b & 1 & & & \\ & 1 & a & 1 & & \\ \cdots & \cdots & \cdots & \cdots & \cdots & \\ & & & 1 & a & 1 \\ & & & & 1 & b \end{bmatrix}, \quad 2n \times 2n.$$

Eigenvalues:

$$\lambda_k = \frac{a + b \pm \left[(a-b)^2 + 16\cos^2 \frac{k\pi}{2n+1}\right]^{\frac{1}{2}}}{2}, \quad k = 1, 2, \ldots, n.$$

Reference: [18, pp. 31-32].

Example 7.9

$$\begin{bmatrix} a & 1 & & & & \\ 1 & b & 1 & & & \\ & 1 & a & 1 & & \\ \cdots & \cdots & \cdots & \cdots & \cdots & \\ & & & 1 & b & 1 \\ & & & & 1 & a \end{bmatrix}, \quad 2n+1 \times 2n+1 .$$

Eigenvalues:

$$\lambda_k = \frac{a + b \pm \left[(a-b)^2 + 16 \cos^2 \frac{k\pi}{2n+2}\right]^{\frac{1}{2}}}{2}, \quad k = 1, 2, \ldots, n.$$

$$\lambda_{2n+1} = a$$

Reference: [18, p. 33].

Example 7.10

Let A_{n+1} be the n+1 x n+1 matrix

$$A_{n+1} = \begin{bmatrix} 0 & x_1 & & & & & \\ y_1 & 0 & x_2 & & & & \\ & y_2 & 0 & x_3 & & & \\ & & \cdots & \cdots & \cdots & & \\ & & & & y_{n-1} & 0 & x_n \\ & & & & & y_n & 0 \end{bmatrix}$$

with $x_k y_k = k(n-k+1)$, $k = 1, 2, \ldots, n$.

Eigenvalues:

If n is an odd integer, the eigenvalues of A_{n+1} are

$$\pm n, \pm(n-2), \ldots, \pm 1.$$

If n is an even integer, the eigenvalues of A_{n+1} are

$$\pm n, \pm(n-2), \ldots, \pm 2, 0.$$

Examples:

A. $x_k = k$, $k = 1, 2, \ldots, n$.

$y_k = n - k + 1$, $k = 1, 2, \ldots, n$.

B. $x_k = i[k(n-k+1)]^{\frac{1}{2}}$, $k = 1, 2, \ldots, n$.

$y_k = -i[k(n-k+1)]^{\frac{1}{2}}$, $k = 1, 2, \ldots, n$.

Inverse: Example 3.19

Reference: [12].

Example 7.11

Let $A = [a_{ij}]$ be the n+1 x n+1 matrix given by

$$A = \begin{bmatrix} -n & n+s & 0 & 0 & \cdots \\ n & -(3n+s-2) & 2(n+s-1) & 0 & \cdots \\ 0 & 2(n-1) & -(5n+2s-8) & 3(n+s-2) & \cdots \\ 0 & 0 & 3(n-2) & -(7n+3s-18) & \cdots \\ \vdots & \vdots & \vdots & \vdots & \end{bmatrix}$$

where s is an arbitrary parameter. In general,

$$a_{ii} = -[(2i+1)n + is - 2i^2]$$

$$a_{i,i+1} = (i+1)(n+s-i)$$

$$a_{i,i-1} = i(n-i+1)$$

$$a_{ij} = 0, \text{ if } |i-j| > 1,$$

where $i,j = 0, 1, 2, \ldots, n$.

Eigenvalues:

$$\lambda_j = -j(s+j+1), \quad j = 0, 1, 2, \ldots, n.$$

Left Eigenvectors:

Let $y^{(j)}$ be the left eigenvector of A corresponding to the eigenvalue λ_j, $j = 0, 1, \ldots, n$. Then the components of $y^{(j)}$ are given by

$$y_i^{(j)} = \frac{1}{\binom{n}{i}} \sum_{k=0}^{q} (-1)^k \binom{n-k}{n-i} \binom{j}{k} \binom{s+j+k}{k}$$

for $i = 0, 1, 2, \ldots, n$, and $q = \min(i,j)$.

Right Eigenvectors:

Let $x^{(j)}$ be the right eigenvector of A corresponding to the eigenvalue λ_j, j = 0, 1, ..., n. Let r be an integer such that $1 \leq r \leq n$. The components of $x^{(j)}$ are

$$x_i^{(j)} = \binom{n+s-i}{n-i} y_i^{(j)} \qquad \text{unless } s = -r \text{ and } j \geq r.$$

If s = -r and j ≥ r, then

$$x_i^{(j)} = \left[\frac{\binom{n+s-i}{n-i}}{(r+s)}\right] y_i^{(j)} \qquad \text{if } i \leq n - r,$$

$$x_i^{(j)} = \left[\frac{\binom{n+s-i}{n-i}}{(r+s)}\right] \frac{y_i^{(j)}}{(r+s)} \qquad \text{if } i > n - r.$$

Reference: [16], [60, pp. 156-157].

REFERENCES

1. Aergerter, M. J. "Construction of a Set of Test Matrices," Comm. ACM, 2, 8 (1959), 10-12.

2. Bellman, Richard Ernest. Introduction to Matrix Analysis, McGraw Hill, New York, 1960.

3. Bodewig, E. "A Practical Refutation of the Iteration Method for the Algebraic Eigenvalue Problem," MTAC, 8 (1954), 237-240.

4. Booth, A. D. Numerical Methods, Butterworths, London, 1957.

5. Boothroyd, J. "Ordering the Roots of Real Symmetric Matrices by Jacobi's Method," submitted to Comm. ACM.

6. Brenner, J. L. "Mahler Matrices and the Equation $QA = AQ^m$," Duke Math. Jour., 29 (1962), 13-28.

7. Brenner, J. L. "A Set of Test Matrices for Testing Computer Programs," Comm. ACM, 5 (1962), 443-444.

8. Brenner, J. L., and G.W. Reitweisner. "Remark on Determination of Characteristic Roots by Iteration," MATC, 9 (1955), 117-118.

9. Brooker, R. A., and F. H. Sumner. "The Method of Lanczos for Calculating the Characteristic Roots and Vectors of a Real Symmetric Matrix," Proc. IEE, 103, pt. 3, supp. 1 (1956), 114.

10. Burgoyne, F. D. "Note 3107, Inverse of a Tridiagonal Matrix," Mathematical Gazette, 48 (1964), 436-437.

11. Caffrey, John. "Another Test Matrix for Determinants and Inverses," Comm. ACM, 6 (1963), 310.

12. Clement, Paul A. "A Class of Triple-diagonal Matrices for Test Purposes," SIAM Review, 1 (1959), 50-52.

13. Cline, Randall E. "A Class of Matrices to Test Inversion Procedures," Comm. ACM, 7 (1964), 725.

14. Dolph, C. L., and D. C. Lewis. "On the Application of Infinite Systems of Ordinary Differential Equations to Perturbations of Plane Poiseuille Flow," Quart. Applied Math., 16 (1958), 97-110.

15. Eberlein, P. J., "A Jacobi-like Method for the Automatic Computation of Eigenvalues and Eigenvectors," Jour. SIAM, 10 (1962), 74-88.

16. Eberlein, P. J. "A Two-parameter Test Matrix," Math. Comp., 18 (1964), 296-298.

17. Eberlein, P. J., and John Boothroyd. "Solution to the Eigenproblem by a Norm Reducing Jacobi Type Method," Num. Math., 11 (1968), 1-12.

18. Elliott, Joseph F. "The Characteristic Roots of Certain Real Symmetric Matrices," Master's thesis, Univ. of Tennessee, 1953.

19. Fairthorne, R. A , and J. C. P. Miller. "Hilbert's Double Series Theorem and Principal Latent Roots of the Resulting Matrix," MTAC, 3 (1949), 399-400.

20. Fettis, Henry E., and James C. Caslin. "Eigenvalues and Eigenvectors of Hilbert Matrices of Order 3 Through 10," Math. Comp., 21 (1967), 431-441.

21. Fiedler, Miroslav. "Some Estimates of the Proper Values of Matrices," Jour. SIAM, 13 (1965), 1-5.

22. Fox, L. An Introduction to Numerical Linear Algebra, Oxford Univ. Press, New York, 1965.

23. Fox, L. "A Short Account of Relaxation Methods," Quart. Jour. Meth. Applied Math., 1 (1948), 253-280.

24. Frank, Werner L. "Computing Eigenvalues of Complex Matrices by Determinant Evaluation and by Methods of Danilewski and Wielandt," Jour. SIAM, 6 (1958), 378-392.

25. Friedman, B. "Eigenvalues of Composite Matrices," Proc. Camb. Phil. Soc., 57 (1961), 37-49.

26. Greenstadt, John. "Some Numerical Experiments in Triangularizing Matrices," Num. Math., 4 (1962), 187-195.

27. Greenwood, Robert. Private correspondence.

28. Hohn, Franz Edward. Elementary Matrix Algebra, The Macmillan Company, New York, 1958.

29. Lanczos, C. Applied Analysis, Prentice-Hall, Englewood Cliffs, New Jersey, 1961.

30. Lehmer, D. H. "Mahler's Matrices," Jour. Aust. Math. Soc., 1 (1959-60), 385-395.

31. Lietzke, M. H., R. W. Stoughton, and Marjorie P. Lietzke. "A Comparison of Several Methods for Inverting Large Symmetric Positive Definite Matrices," Math. Comp., 18 (1964), 449-456.

32. Lotkin, Mark. "Determination of Characteristic Values," Quart. Applied Math., 17 (1959), 237-244.

33. Lotkin, Mark. "A Set of Test Matrices," MTAC, 9 (1955), 153-161.

34. Macon, N., and A. Spitzbart. "Inverses of Vandermonde Matrices," Amer. Math. Monthly, 65 (1958), 95-100.

35. Mahler, K. "A Representation of the Primitive Residue Classes (mod 2n)," Proc. Amer. Math. Soc., 8 (1957), 525-531.

36. Marcus, Marvin. Basic Theorems in Matrix Theory, Nat. Bur. Standards A. M S. No. 57, 1960.

37. Martin, R. S., C. Reinsch, and J. H. Wilkinson. "Householder's Tridiagonalization of a Symmetric Matrix," Num. Math. 11 (1968), 181-195.

38. Martin, R. S., and J. H. Wilkinson. "Reduction of the Symmetric Eigenproblem $Ax = \lambda Bx$ and Related Problems to Standard Form," Num. Math., 11 (1968), 99-110.

39. Mueller, Dennis J. "Householder's Method for Complex Matrices and Eigensystems of Hermitian Matrices," Num. Math., 8 (1966), 72-92.

40. Nering, Evar D. Linear Algebra and Matrix Theory, John Wiley and Sons, London, 1963.

41. Newberry, A. C. R. "A Family of Test Matrices," Comm. ACM, 7 (1964), 724.

42. Newman, Morris. "Matrix Computations," in Survey of Numerical Analysis, John Todd, ed., McGraw-Hill, New York, 1962.

43. Newman, Morris, and John Todd. "The Evaluation of Matrix Inversion Programs," Jour. SIAM, 6 (1958), 466-476.

44. Ortega, James M. "Generation of Test Matrices by Similarity Transformations," Comm. ACM, 7 (1964), 377-378.

45. Parlett, Beresford. "Laguerre's Method Applied to the Matrix Eigenvalue Problem," MTAC, 18 (1964), 464-485.

46. Pei, M. L. "A Test Matrix for Inversion Procedures," Comm. ACM, 5 (1962), 508.

47. Rosser, J. B., C. Lanczos, M. R. Hestenes, and W. Karush. "Separation of Close Eigenvalues of a Real Symmetric Matrix," Jour. Res. Nat. Bur. Standards, 47 (1951), 291-297.

48. Ruhe, A. "On the Quadratic Convergence of a Generalization of the Jacobi Method to Arbitrary Matrices," BIT, 8 (1968), 210-231.

49. Rutishauser, Heinz. "Solution of Eigenvalue Problems with the LR-Transformation," in Further Contributions to the Solution of Simultanous Linear Equations and the Determination of Eigenvalues, Nat. Bur. Standards A. M. S. No. 49 (1958), 47-81.

50. Savage, Richard, and Eugene Lukacs. "Tables of Inverses of Finite Segments of the Hilbert Matrix," in Contributions to the Solution of Systems of Linear Equations and the Determination of Eigenvalues, Nat. Bur. Standards A. M. S. No. 39 (1954), 105-108.

51. Semendiaev, K. A. "The Determination of Latent Roots and Invariant Manifolds of Matrices by Means of Iterations," Nat. Bur. Standards Report 1402.

52. Snyder, James N. "Improvement of the Solutions to a Set of Simultaneous Linear Equations Using the ILLIAC," MTAC, 9 (1955), 177-184.

53. Taussky, Olga, and John Todd. "Systems of Equations, Matrices, and Determinants," Math. Mag., 26 (1952), 71-88.

54. Thompson, Gene. "Characteristic Values and Vectors of Defective Matrices," Comm. ACM, 6 (1963), 106-107.

55. Todd, John. "The Condition of Finite Segments of the Hilbert Matrix," in Contributions to the Solution of Systems of Linear Equations and the Determination of Eigenvalues, Nat. Bur. Standards A. M. S. No. 39 (1954), 109-116.

56. Todd, John. UCLA Extension Course Lecture, June 16, 1948.

57. Turing, A. M. "Rounding-off Errors in Matrix Processes," Quart. Jour. Mech. Applied Math., 1 (1948), 287-308.

58. Voigt, Susan. Private correspondence.

59. Von Neumann, J., and H. H. Goldstine. "Numerical Inverting of Matrices of High Order," Bull. Amer. Math. Soc., 53 (1947), 1021-1099.

60. Westlake, Joan R. A Handbook of Numerical Matrix Inversion and Solution of Linear Equations, John Wiley and Sons, Inc., New York, 1968.

61. White, Paul A. "The Computation of Eigenvalues and Eigenvectors of a Matrix," Jour. SIAM, 6 (1958), 393-437.

62. Wilkinson, J. H. The Algebraic Eigenvalue Problem, Clarendon Press, Oxford, 1965.

63. Wilkinson, J. H. "The Calculation of the Eigenvectors of Codiagonal Matrices," Comp. Jour., 1 (1958), 90-96.

64. Wilkinson, J. H. "Error Analysis of Direct Methods of Matrix Inversion," Jour. ACM, 8 (1961), 281-330.

65. Wilkinson, J. H. "Error Analysis of Floating-point Computation," Num. Math., 2 (1960), 319-340.

66. Wilkinson, J. H. "The Evaluation of Zeros of Ill-conditioned Polynomials, Part II," Num. Math., 1 (1959), 167-180.

67. Wilkinson, J H. "Householder's Method for the Solution of the Algebraic Eigenproblem," Comp. Jour., 3 (1960), 23-27.

68. Wilkinson, J. H. "Householder's Method for Symmetric Matrices," Num. Math., 4 (1962), 354-361.

69. Wilkinson, J. H. "Instability of the Elimination Method of Reducing a Matrix to Tridiagonal Form," Comp. Jour., 5 (1962), 61-70.

70. Wilkinson, J. H. Private correspondence.

71. Wilkinson, J. H. "Rigorous Error Bounds for Computed Eigensystems," Comp. Jour., 4 (1961), 230-241.

72. Wilkinson, J. H. Rounding Errors in Algebraic Processes, Prentice-Hall, Inc., Englewood Cliffs, New Jersey, 1963.

73. Wilkinson, J. H. "Stability of the Reduction of a Matrix to Almost Triangular and Triangular Forms by Elementary Similarity Transformations," Jour. ACM, 6 (1959), 336-359.

74. Wilkinson, J. H. "The Use of Iterative Methods for Finding the Latent Roots and Vectors of Matrices," MTAC, 9 (1955), 184-191.

* * * * * *

75. Milnes, Harold, Willis. "A Note Concerning the Properties of a Certain Class of Test Matrices," Math. Comp., 22 (1968), 827-832.

76. Givens, J. W. "Conference on Matrix Computations," Jour. ACM, 4 (1958), 20.

77. Todd, John. "The Problem of Error in Digital Computation," in Error in Digital Computation, Vol. I, L. B. Rall, ed., Wiley, New York, 1965.

78. Forsythe, George E. Private correspondence.

79. Varah, James. "The Computation of Bounds for the Invariant Subspaces of a General Matrix Operator," Report CS66 (1967), Computer Science Department, Stanford University.

80. Gregory, Robert T. "Defective and Derogatory Matrices," Siam Review, 2 (1960), 134-139.

81. Knuth, D. E. Fundamental Algorithms (Volume I in the series The Art of Computer Programming), Addison-Wesley, Reading, Mass., 1968.

SYMBOL TABLE

Symbol	Meaning
A, A_n	an n x n matrix
$[a_{ij}]$	an n x n matrix whose i,j element is a_{ij}
A_{ij}	the i,j submatrix of a partitioned matrix
$\bar{A}$	the complex conjugate of A
A^H	the complex conjugate transpose of A
A^T	the transpose of A
A^{-1}	the inverse of A
$\|A\|$	the norm of A
$A^{(k)}$	(i) the kth matrix in a sequence (ii) the kth Kronecker power of A
$A \otimes B$	the Kronecker product of A and B
$\det(A)$, $\|A\|$	the determinant of A
$K(A)$	general condition number of A
$M(A)$	Turing's M-condition number
$N(A)$	Turing's N-condition number
$P(A)$	(i) von Neumann and Goldstine's condition number (ii) a polynomial in A
I	the identity matrix of order n
I_m	the identity matrix of order m
x	an n-dimensional column vector

x^T	the corresponding row vector (x transpose)
x^H	the complex conjugate transpose of x
(x,y), $y^H x$	the scalar product of x and y
xy^H	a square matrix (do not confuse with $y^H x$)
$x^{(j)}$	the jth vector in a sequence
$x_i^{(j)}$	the ith component of $x^{(j)}$
λ	an eigenvalue
λ_i	the ith eigenvalue
$P(\lambda)$	the characteristic polynomial
s_i	the condition number of λ_i
(k,n)	the greatest common divisor of k and n
$(k,n) = 1$	k and n are relatively prime
$\exp(r)$	e^r
ω	cube root of unity
δ_{ij}	the Kronecker delta = 1, if $i = j$, 0, otherwise
J_{nk}	$n \times k$ matrix, each element is 1
f_n	the n-dimensional vector, each component is 1

e_n^i	column vector of dimension n whose components are δ_{ij}, $j = 1,2,\ldots,n$
g_i	$f_n - ne_n^i \quad i = 1,2,\ldots,n-1$
$a \equiv b \pmod{m}$	$a - b$ is divisible by m
$a \doteq b$	a is approximately equal to b
$a_n \sim b_n$	a_n is asymptotic to b_n
$\sim$	equivalence relation
$\binom{n}{r}$	binomial coefficient
$a \mid b$	a divides b
$\lvert a \rvert$	absolute value of a
i	$\sqrt{-1}$

INDEX